THE AFRICAN NEOGENE –
CLIMATE, ENVIRONMENTS AND PEOPLE

Palaeoecology of Africa

International Yearbook of Landscape Evolution and Palaeoenvironments

ISSN 2372-5907

Volume 34

Editor in Chief

J. Runge, Frankfurt, Germany

The African Neogene – Climate, Environments and People

Editor

Jürgen Runge
Centre for Interdisciplinary Research on Africa (CIRA/ZIAF),
Johann Wolfgang Goethe University, Frankfurt am Main, Germany

In collaboration with

Joachim Eisenberg
Institute of Physical Geography, Johann Wolfgang Goethe University,
Frankfurt am Main, Germany

CRC Press
Taylor & Francis Group
Boca Raton London New York Leiden

CRC Press is an imprint of the
Taylor & Francis Group, an **informa** business

A BALKEMA BOOK

Front cover: Togolese Geographer Tignoati Kolani investigates laminated alluvial sediments of the Keran River, a tributary of the Oti-Volta catchment in West Africa (Photo: J. Runge, April 2016).

CRC Press/Balkema is an imprint of the Taylor & Francis Group, an informa business

© 2018 Taylor & Francis Group, London, UK

Typeset by V Publishing Solutions Pvt Ltd., Chennai, India

Library of Congress Cataloging-in-Publication Data

Applied for

Published by: CRC Press/Balkema
 Schipholweg 107C, 2316 XC Leiden, The Netherlands
 e-mail: Pub.NL@taylorandfrancis.com
 www.crcpress.com – www.taylorandfrancis.com

First issued in paperback 2020

ISBN 13: 978-0-367-57291-4 (pbk)
ISBN 13: 978-1-138-06212-2 (hbk)

Contents

Foreword

The eleven papers presented in the 34th volume of Palaeoecology of Africa (PoA) illustrate both the ongoing challenges and problems facing Neogene and Quaternary environmental reconstruction on the African continent. They look at the climate and environments of the past at different temporal and spatial scales as well as the role played by humans in those environments. Methodologically, the contributions cover a wide range of interdisciplinary approaches such as the geomorphological shape of landscapes (macro-scale river basin evolution); interpretation of sedimentary records of deltas, caves and rock-shelters; and the palynological significance of pollen grains from high altitude floating bogs. Interpretation of these findings to correctly reconstruct ancient landscapes' ecosystem processes is always complex and a challenge for the authors.

In this volume, a strong regional focus is set on Nigeria in order to reflect the basic research carried out by growing number of African scientists, which is often linked to applied research supporting the oil and gas industry in the Niger Delta. Besides near-coast Nigerian case studies, regional research in this book covers the margins of the wider Congo Basin, for example, Loango Bay (R. Congo), the eastern Kivu Rift Valley (D.R. Congo), and the Mumba Rock Shelter (Tanzania). One applied contribution from West Africa looks at the growing importance of Developing Minerals and GIS techniques in the urban areas of Accra (Ghana). The role of previous research carried out in the Tibesti Mountains in Central Sahara is recalled by a paper, which also highlights future research potential that will contribute to the not-yet-answered questions of the African palaeoenvironment.

Two obituaries pay tribute to longstanding members of the PoA Editorial Board: the late colleagues Karl W. Butzer (1934–2016) and Françoise Gasse (1942–2014).

The peer review process was thankfully supported by numerous anonymous reviewers. Two student assistants, Tobias Buchwald and Nishtha Prakash, from the University of Frankfurt Physical Geography Working Group and the Centre for Interdisciplinary Research on Africa (ZIAF) helped streamline the editing process. Dr. Joachim Eisenberg reliably formatted the manuscripts for PoA layout and style. He also revised many figures and carried out additional and complex cartographic art work for this book.

Finally, I take this opportunity to place on record my gratitude to Senior Publisher Janjaap Blom and his team from Routledge/Taylor & Francis/CRC Press in Leiden (The Netherlands) for the continuous support to PoA.

Jürgen Runge
Frankfurt
June 2017

Contributors

Olabisi Adekeye

Department of Geology and Mineral Science, University of Ilorin, Nigeria

Onema Adojoh

School of Environmental Sciences, University of Liverpool, L69 7ZT, United Kingdom; Geosciences and Geological and Petroleum Engineering, University of Missouri Science and Technology, Rolla, MO 65409-0410, USA, Email: adojoho@mst.edu

Mike I. Akaegbobi†

Department of Geology, University of Ibadan, Nigeria

Samuel Akande

Department of Geology and Mineral Science, University of Ilorin, Nigeria

Akintunde Akintola

Geology Department, Olabisi Onabanjo University, Ago-Iwoye, Nigeria

Samson Bankole

Department of Geosciences, University of Lagos, Nigeria, Email: sbankole@unilag.edu.ng

Olugbenga A. Boboye

Department of Geology, University of Ibadan, Ibadan, Nigeria, Email: boboyegbenga@yahoo.com

Florent Boudzoumou

Département de Géologie, Université Marien Ngouabi, Brazzaville, Congo

Pastory G.M. Bushozi

Department of Archaeology and Heritage University of Dar es Salaam, Tanzania, Email: pbushozi@gmail.com

Jacinta N. Chukwuma-Orji

Department of Geology, Federal University of Technology, Minna, Nigeria, Email: jacinta@futminna.edu.ng

Woody (Fenton) P.D. Cotterill

Geoecodynamics Research Hub, c/o Department of Earth Sciences, University of Stellenbosch, Matieland, 7602, South Africa, Email: fenton.cotterill@uct.ac.za

Olusola Dublin-Green

Department of Geosciences, University of Lagos, Nigeria

Robert Duller

School of Environmental Sciences, University of Liverpool, L69 7ZT, United Kingdom

Frank D. Eckardt

Department of Environmental and Geographical Science, University of Cape Town, Rondebosch, 7701, South Africa, Email: frank.eckardt@uct.ac.za

Dupe Egbeola

Department of Geology, University of Ibadan, Ibadan, Nigeria

Bernd-D. Erdtmann

Institut für Angewandte Geowissenschaften, Technische Universität Berlin, Sekr. EB 10, Ernst-Reuter-Platz 1, D-10587, Berlin, Germany

Tyrel J. Flügel

Department of Environmental and Geographical Science, University of Cape Town, Rondebosch, 7701, South Africa; Department of Environmental and Geographical Science, University of Cape Town, Rondebosch, 7701, South Africa, Email: tyrel.flugel@gmail.com

Baldur Gabriel

Gertraudenstraße 13, 16225 Eberswalde, Germany, Email: Baldur.Gabriel@web.de

Pierre Giresse

Centre de Formation et de Recherches sur les Environnements Méditerranéens, Université de Perpignan, Perpignan, France, Email: giresse@univ-perp.fr

Isah A. Goro

Department of Geology, Federal University of Technology, Minna, Nigeria

Chantal Kabonyi Nzabandora

Faculty of Sciences, University of Bukavu, D.R. Congo, Email: chantalkabonyi@gmail.com

Luis Leque

Natural Museum, Madrid, Spain

Audax Mabulla

National Museums of Tanzania, Tanzania

Dieudonné Malounguila-Nganga

Département de Géologie, Université Marien Ngouabi, Brazzaville, Congo, Email: malounguila@hotmail.fr

Fabienne Marret

School of Environmental Sciences, University of Liverpool, L69 7ZT, United Kingdom, Email: F.Marret@liverpool.ac.uk

Timothée Miyouna

Département de Géologie, Université Marien Ngouabi, Brazzaville, Congo,
Email: miyounatim@yahoo.fr

Rosemary Okla

Ghana Geological Survey Authority, Accra, Ghana,
Email: rosemaryokla@gmail.com

Edward A. Okosun

Department of Geology, Federal University of Technology, Minna, Nigeria

Samuel Olobaniyi

Department of Geosciences, University of Lagos, Nigeria

Peter Osterloff

Shell UK Limited, Aberdeen, AB12 3FY, United Kingdom,
Email: Peter.Osterloff@shell.com

Emile Roche

Palaeontology, University of Liège, Belgium,
Email: rocheemile@yahoo.fr

Jürgen Runge

Centre for Interdisciplinary Research on Africa (ZIAF) and Institute of Physical
Geography, Johann Wolfgang Goethe University, Altenhöferallee 1, 60438
Frankfurt am Main, Germany, Email: j.runge@em.uni-frankfurt.de

Eckert Schrank

Institut für Angewandte Geowissenschaften, Technische Universität
Berlin, Sekr. EB 10, Ernst-Reuter-Platz 1, D-10587, Berlin, Germany

Florence Sylvestre

CEREGE, Technopole Environnement Arbois-Mediterranée,
Aix-en-Provence, France, Email: sylvestre@cerege.fr

Salome H. Waziri

Department of Geology, Federal University of Technology, Minna, Nigeria

CHAPTER 1

Obituaries: Karl W. Butzer (1934–2016)
Françoise Gasse (1942–2014)

Jürgen Runge

Centre for Interdisciplinary Research on Africa (CIRA/ZIAF),
Johann Wolfgang Goethe University, Frankfurt am Main, Germany

Florence Sylvestre

CEREGE, Technopole Environnement Arbois-Mediterranée,
Aix-en-Provence, France

It is with deep regret to report that 'Palaeoecology of Africa (PoA)' lost two of its long-time editorial board members: the German-born, US-American geographer, palaeoecologist and archaeologist, Karl W. Butzer in 2016; and the French palaeo-biologist (diatom specialist), palaeoclimatologist and palaeohydrologist, Françoise Gasse in 2014. Both supported and accompanied over many years the long-term success story of PoA. Founded in 1966 by Professor Eduard M. van Zinderen Bakker (1907–2002) and assisted by Professor Joey Coetzee (1921–2007, see Palaeoecology of Africa, Vol. 29, 2009) the PoA series has consistently published interdisciplinary scientific papers on landscape evolution and on former environments of the African continent as well as papers on changes in climate and in vegetation cover interconnected to environmental dynamics from the Cainozoic up to the present. Recently, the PoA also broadened its horizons to the steadily growing influence of humans on many of the field sites studied that has shaped the scientific profile of the series.

1.1 KARL W. BUTZER

Karl W. Butzer (Figure 1) was born on August 19, 1934 in Mühlheim an der Ruhr (Rhineland, Germany). The rise of Nazi Germany coincided with his childhood, and in 1937, his Catholic family decided to escape from Germany. Karl and his older brother Paul were smuggled under the seat of a school bus to England, where they later reunited with their parents. Finally, during World War II in 1941, the Butzers emigrated to Montreal, Canada (Doolittle, 2016; Turner II, 2017). "The trauma of migration, family separation, persecution, […] incidents of prejudice in Canada, and an uncertain fate of family members left behind, are all experiences that have influenced [Karl Butzer's] outlook on life" (Offen, 2003, p. 125).

Subsequently he studied at McGill University (Montreal) where he received two degrees—the B.Sc. (Honours) in Mathematics (1954), and the M.Sc. in Meteorology and Geography (1955). With the assistance of an exchange fellowship, he returned to Germany and studied under Carl Troll (1899–1975), the initiator of the concept

Figure 1. Karl W. Butzer in 2005 (Source: Sounny Wikimedia commons licence).

of "landscape ecology", at the University of Bonn. In 1957, shortly before his 23rd birthday, he received the doctorate in natural science (Dr. rer. nat.) in the disciplines of Physical Geography and Ancient History. By graduation, Butzer already had published seven scientific articles!

In Bonn, Karl Butzer met Elisabeth Schlösser who became his wife and collaborator for 56 years. "The two honeymooned on Mallorca, where they started a not-so-romantic project on palaeosols and fossil beaches" (Doolittle, 2016). Subsequently the couple worked jointly together in Egypt, Ethiopia, and South Africa. "On their numerous projects in Spain and Mexico, Karl excavated and mapped, while Elisabeth ('Lis') scoured archives and translated documents" (Doolittle, 2016).

His outstanding performance at an early stage of his career allowed him to work for two years as a research associate at the German Academy of Sciences and Literature in Mainz, Germany (Turner II, 2017). He was then Assistant, and later Associate Professor at the University of Wisconsin-Madison (1959–1966).

In 1966, he accepted an offer as Professor of Anthropology and Geography at the University of Chicago where he was named the "Henry Schultz Professor of Environmental Archaeology" in 1980. He was elected to various sub-departments, namely the Committee on African Studies, Committee on Evolutionary Biology, Committee on Archaeological Studies (Humanities), and as Professor in the Oriental Institute (University of Texas, 2016).

"Pleistocene geology is primarily concerned with stratigraphy and chronology. A more comprehensive study of past environments is needed, a Pleistocene Geography concerned with the natural environment and focused on the same themes of "man and nature" that are the concern of historical and contemporary geographies. This is a field to which both the natural scientists and the archaeologist should contribute—more directly and with greater enthusiasm" (Butzer, 1964, p. 7)

The University of Chicago provided up to the year 1984 the ideal intellectual home for Karl Butzer (Doolittle, 2016). He viewed Chicago as a great place to live and to raise four children. Butzer's geoarchaeological research was much appreciated by his colleagues, with whom he not only collaborated, but also considered them as lifelong friends. During the Chicago years, he wrote landmark books such as 'Desert and River in Nubia: Geomorphology and Prehistoric Environments at the Aswan Reservoir (1968)' (with C. L. Hansen), 'Early Hydraulic Civilization in Egypt: A Study in Cultural Ecology (1976)', 'Geomorphology from the Earth (1976)', and 'Archaeology as Human Ecology (1982): Method and Theory for a Contextual Approach' (Turner II, 2017). It was obvious that Butzer's thematic approach and scientific research was, at an early stage, already focused on 'inter- and transdisciplinary' perspectives, and therefore, he expanded his efforts on the social dimensions of human-environment relationships.

In 1981, Butzer was visiting Chair and Professor of Human Geography at the Swiss Federal Institute of Technology (ETH) in Zurich for one year. In Zurich, he advanced a full-fledged human-environment science of the past in Archaeology as Human Ecology. Butzer amplified his challenge to palaeoenvironmental and archaeological researchers to appreciate the complexity of human-environment interactions and to allow the exploration of a full range of evidence. Subsequently, from 1982 until 1987, Karl Butzer and his wife carried out comprehensive studies in eastern Spain, combining archaeological excavation, archival documentation as well as settlement and land use studies (Turner II, 2017).

In 1984, Butzer took his final position as the Raymond C. Dickson Professor of Liberal Arts at the Departments of Geography and Anthropology, University of Texas at Austin. In driving proximity to Mexico, Karl Butzer and his wife undertook extensive excursions and field work in the Bajio of north-central Mexico, addressing the environmental impacts of Hispanic land uses. They mainly focused on archival documentation (Lis) and on sediments (Karl) (Turner II, 2017). This research challenged claims about the environmentally degrading land uses of the Colonial period, noting that numerous indicators of pre-Hispanic land degradation were prevalent in the Bajio, some of which improved under Hispanic dominion. As Butzer was always insisting on an evidence-based interpretation of land changes in Mexico, he challenged the popular and common position accompanying the Columbus quincentenary that the Spanish introduction of herd animals and the plough were the fundamental elements for land degradation within colonial Mexico (Turner II, 2017). For his comprehensive contributions to the geography of Latin America, in 2002 he received the Preston E. James Eminent Latin Americanist Career Award (Offen, 2003).

Population cycles and civilizational collapse, including articles on institutional structures, demography, climatic forcing, and environmental degradation appeared in 1980–1982, 1990, 1994, and 1997. Butzer organized a symposium on 'Collapse, Environment and Society' in 2010, and a review presentation appeared in the Proceedings of the National Academy of Sciences, in conjunction with a set of specialist papers, co-edited with Georgina Endfield in 2012 (University of Texas, 2016).

Other themes of Karl Butzer addressed in terms of geoscience included studies on coastal geomorphology (Egypt, Spain, South Africa, and Atlantic Canada, between 1960 and 2002), and tufa waterfalls, playa lakes, or periglacial phenomena in South Africa and Spain (between 1964 and 1979). Biographical themes bring published recollections on emigration, ethnic prejudice, and academic freedom in the authoritarian state (2001–2004) (University of Texas, 2016).

His last paper in 2016 on 'Dry lakes or *Pans* of the Western Free State, South Africa: Environmental history of Deelpan and possible early human impacts' was published in honour of Professor Louis Scott, Bloemfontein, South Africa, in the series *Palaeoecology of Africa* (Volume 33).

Karl Butzer served as editor from 1978–2009 for the *Journal of Archaeological Science*, and as series editor of *Prehistoric Archaeology and Ecology* for the University of Chicago Press (1973–1988). He also held positions as member of the editorial boards of many distinguished journals such as *Advances in Archaeological Method and Theory, Catena, Geographical Review, Geomorphology, Physical Geography, Progress in Human Geography, Palaeocology of Africa, Paleorient, Stratigraphic Newletters,* and *Quaternaria.*

"Following the traditions of natural history, Karl Butzer believed that both inductive and deductive approaches, always informed by empirical evidence, were required to arrive at robust interpretations. He was suspicious of purely theoretically driven research applied to past and present human-environment interactions, commonly responding by demonstrating instances in which the evidence ran counter to theory or, at least, required a much more nuanced understanding of the processes in question. This understanding was not fully gained by the mechanistic explications of the natural sciences, but required an appreciation of what historical, ethnographic, and cultural evidence had to offer" (Turner II, 2017, p. 5).

Karl Butzer loved field research, presenting his results to the scientific community, teaching and mentoring, especially at the graduate level. He received the Graduate Teaching Award from the University of Texas at Austin in 2005 (University of Texas, 2016). His research output was remarkable. He published 14 books and monographs, 275 refereed articles, book chapters, and encyclopaedia entries written in English, German, and Spanish; some having been translated into six other languages. He was most at home in the field, and his field of research was extensive! (Turner II, 2017). This is appropriately underlined by the following quote: "… his students in geography, archaeology, geology, and Latin American, African, and Middle Eastern Studies attest, you could learn more in one day in the field with Professor Butzer than you could in a semester-long course with any other professor…" (Doolittle, 2016).

On May 4, 2016, at the age of 81 Karl W. Butzer passed away in Austin, Texas after a short illness (Knapp 2016). "Karl is survived by Lis, two daughters, Helga and Kieke, and two sons, Carl and Hans, seven grandchildren, and his older brother, Paul" (Turner II, 2017, p. 7). He will be greatly missed.

1.1.1 Selected bibliography of Karl W. Butzer

1960   Archaeology and geology in ancient Egypt. *Science*, **132**, pp. 1617–1624.
1964   *Environment and Archaeology: An Introduction to Pleistocene Geography*, (Chicago: Aldine).
1965   Acheulian occupation sites at Torralba and Ambrona, Spain: their geology. *Science*, **150**, pp. 1718–1722.
1968   with C.L. Hansen, *Desert and River in Nubia: Geomorphology and Prehistoric Environments at the Aswan Reservoir*, (Madison: University of Wisconsin Press).
1969   Changes in the Land. *Science*, **165**, pp. 52–53.
1971   *Environment and Archaeology: An Ecological Approach to Prehistory*, (Chicago: Aldine).

1972   with Isaac, G.L., Richardson, J.L. and Washbourn-Kamau, C., Radiocarbon dating of East African lake levels. *Science*, **175**, pp. 1069–1076.

1973   Paleo-hydrology of late Pleistocene lakes in the Alexandersfontein Pan, Kimberly, South Africa. *Nature*, **243**, pp. 328–330.

1976   *Early Hydraulic Civilization in Egypt: A Study in Cultural Ecology*, (Chicago: University of Chicago Press).

1976   *Geomorphology from the Earth*, (New York: Harper and Row).

1979   with Fock, G.J., Scott, L. and Stuckenrath, R., Dating and context of rock engravings in southern Africa. *Science*, **203**, pp. 1201–1214.

1980   Volcanism in human history. *Science*, **208**, pp. 736–737.

1982   *Archaeology as Human Ecology: Method and Theory for a Contextual Approach*, (Cambridge: Cambridge University Press).

1983   Human Response to Environmental Change in the Perspective of Future, Global Climate. *Quaternary Research*, **19**, pp. 279–292.

1986   with Butzer, E.K. and Mateu, J.F., Medieval Muslim Communities of the Sierra de Espadan, Kingdom of Valencia. *Viator*, **17**, pp. 339–413.

1988   Cattle and sheep from Old to New Spain: Historical antecedents. *Annals of the Association of American Geographers*, **78**, pp. 29–56.

1993   No Eden in the New World. *Nature*, **362**, pp. 15–17.

1996   Ecology in the long view: settlement histories, agrosystemic strategies, and ecological performance. *Journal of Field Archaeology*, **23**, pp. 141–150.

2005   Environmental history in the Mediterranean world: cross-disciplinary investigation of cause-and-effect for degradation and soil erosion. *Journal of Archaeological Science*, **32**, pp. 1773–1800.

2008   Challenges for a cross-disciplinary geoarchaeology: the intersection of environmental history and geomorphology. *Geomorphology*, **101**, pp. 402–411.

2011   Geoarchaeology, climate change, sustainability: A Mediterranean perspective. *The Geological Society of America Special Papers*, **476**, pp. 1–14.

2012   Collapse, environment, and society. *Proceedings of the National Academy of Sciences, USA*, **109**, pp. 3632–3639.

1.1.2 References about Karl W. Butzer

Doolittle, W.E., 2016, Karl W. Butzer: Interdisciplinary mentor. Retrospective. *Proceedings of the National Academy of Sciences of the United States of America*, **113**(41), pp. 11382–11383, www.pnas.org/cgi/doi/10.1073/pnas.1614514113.

Knapp, G., 2016, In Memoriam: Karl Butzer, Latin Americanist Geographer. *Journal of Latin American Geography,* **15**(2), pp. 167–171.

Offen, K.H., 2003, Dr. Karl W. Butzer: Recipient of 2002 Preston E. James Eminent Latin Americanist Award. *Journal of Latin American Geography*, **2**(1), pp. 125–127.

Turner II, B.L., 2017, Karl W. Butzer 1934–2016. *Biographical Memoirs*, pp. 1–9, National Academy of Sciences, www.nasonline.org/memoirs.

University of Texas, 2016, *In Memoriam Dr. Karl W. Butzer*, 3 p., http://sites.utexas.edu/butzer.

1.2 FRANÇOISE GASSE

Françoise Gasse was born on June 22, 1942 in Bergerac (Département Dordogne) in the south-western region of France (Figure 2). Her early childhood was impacted by political uncertainty due to World War II (German occupation) and to post-war experiences of her family and relatives.

Figure 2. Françoise Gasse at work in the laboratories at CEREGE (Gasse 2014, p. 140).

Until much later, she left her home in the French countryside for higher education and went to the greater Paris area. As she was always interested in nature, she oriented herself towards natural sciences and received in 1967 her first university diploma (*Agrégation de Sciences Naturelles*). For almost 20 years, from 1967–1986, she worked as tutor and lecturer in Botany at the traditional higher vocational school for teacher training (*Ecole Normale Supérieure*) in Fontenay-aux-Roses, a suburb south of Paris.

In 1970, while still lecturing at Fontenay-aux-Roses, Françoise Gasse started to work on a dissertation in the Laboratory of Quaternary Geology (Professor Henriette Alimen). The PhD was completed in 1975 when receiving the *Docteur-ès-Sciences, Sciences de la Terre* at the *Université Pierre et Marie Curie* in Paris. Her dissertation on Lake Abhé near the Ethiopia-Djibouti border (Afar) was the first continuously dated African Plio-Pleistocene diatom record. In her dissertation, Gasse showed, through an innovative methodology, that Lake Abhé—today a small, shallow, hyper-alkaline waterbody—was a 160 metre deep freshwater lake that extended over 5000 km^2 during the early to Mid-Holocene (see publication in Nature, Gasse, 1977).

Since 1986, she was permanently employed at the National Centre for Scientific Research (CNRS, *Centre National de la Recherche Scientifique*) in Orsay, Paris, a public organization under the responsibility of the French Ministry of Education and Research, where she was Research Director and was responsible for the bioindicator, climate and environmental modelling team at the hydrological isotope lab. Shortly after the turn of the millennium, Françoise Gasse moved from Orsay to the CEREGE (*Centre Européen de Recherche et d'Enseignement en Géosciences de l'Environnement*) in Aix-en-Provence. She was always very active, both at national and international level, for example as a member of PAGES/IGBP/PEP and the International Continental Scientific Drilling Program (ICDP).

Her involvement in the PAGES (Past Global Changes) programme with regional- to continent-wide landscape transects (Pole-Equator-Pole) through Europe and Africa (PEP3, 1991–2001) sharpened Gasse's interest and understanding for inter-hemispheric teleconnections, ocean-land relationships, climate model reliability, and model-data comparison that were subsequently used to validate models. An important objective of the PEP3 project was to provide quantitative parameters and compare them to simulations derived from climate models. One way to translate palaeolimnological data (lake level, salinity, and isotopic composition) in terms of quantified controlling factors is by hydrological modelling of individual lakes. This approach helps elucidate lake functioning, can yield estimates of past conditions in a lake basin, and can be used to predict lake status under specific future scenarios (Gasse, 2014).

Gasse as a specialist of diatoms initiated pioneer research to reconstruct Quaternary climates and environments in the Sahara and the Sahel, East Africa (Ethiopia), Madagascar, in western and south Asia (Caspian Sea, Tibet), and in the Middle East (Lebanon). Her research commonly integrated diatom and isotopic data. It was characterized both by her sophisticated understanding of the importance of basin hydrogeomorphology in palaeoclimatic interpretation, and the rigour of her taxonomic identification of diatoms.

Therefore, the present knowledge of palaeoclimatology within arid zones is mainly based on her studies on sedimentary archives from lakes and palaeolakes. One of her key contribution has been to develop the use of diatom assemblages to quantify how lake physico-chemistry parameters (salinity, pH) have evolved over time.

Most of her numerous and well-cited articles (>200) deal with the late Quaternary environment, including the late glacial and the last climatic transition. But Gasse's focus was also on the more recent period, the Holocene and the last millennia, with important contributions showing the existence of rapid events—as seen from the hydrological cycle—during those periods that were previously thought to be stable, with major dry spells around 8000 and 4000 years ago.

She was an engaged editorial board member of internationally recognized scientific journals such as *Earth and Planetary Letters, Geology, Nature, Palaeoecology of Africa, Science, Science Reviews, Quaternary Science Reviews*, and many others. She supervised 25 doctoral theses closely linked to her interdisciplinary, multi-proxy data studies.

In 2005, Françoise Gasse was the first woman to receive the Vega Medal in Gold awarded by the Swedish Society for Anthropology and Geography. In 2010, she was awarded the Hans Oeschger Medal of the European Geosciences Union (EGU) for her contribution to the reconstruction of climate variability during the Holocene. Her last contribution to the journal Palaeolimnology in January 2014 was an ultimate tribute to the deserts, "Reminiscences and acknowledgments from a lover of deserts near the end of her professional life" (Gasse, 2014).

"I appreciate having friends throughout the world. I confess, however, that my best 'souvenirs' are and will remain the fantastic memories of travels to remote countries, seeing landscapes and colours that I could not have imagined and having the opportunities to interact with fascinating cultures and observe so many different ways of life" (Gasse, 2014, p. 143).

Francoise Gasse passed away on April 22, 2014. She was a scientific pioneer on many fronts. Her impact is ongoing and the quality of her career is exemplary. Her friendly and discrete authority, her radiant smile, and her cleverness will remain in our memories.

1.2.1 Selected bibliography of Françoise Gasse

1977   Evolution of Lake Abhé (Ethiopia and T.F.A.I.) from 70.000 B.P., *Nature*, **2**, pp. 42–45.
1980   with Rognon, P. and Street, F.A., Quaternary history of the Afar and Ethiopian Rift lakes. In *The Sahara and the Nile*, edited by Williams, M.A.J. and Faure, H., (Rotterdam: Balkema), pp. 361–400.
1986   *East African diatoms: Taxonomy, ecological distribution*, (Berlin: Cramer).
1989   with Ledée, V., Massault, M. and Fontes, J.-C., Water-level fluctuations of Lake Tanganyika in phase with oceanic changes during the last glaciation and deglaciation. *Nature*, **342**, pp. 57–59.
1990   with Téhet, R., Durand, A., Gibert, E. and Fontes, J.-C., The arid-humid transition in the Sahara and the Sahel during the last deglaciation. *Nature*, **346**, pp. 141–146.
1991   with Arnold, M., Fontes, J.C., Fort, M., Gibert, E., Huc, A., Binyan, L., Yuanfanfg, L., Mélières, F., van Campo, E., Fubao, W. and Qingsong, Z., A 13000–Year climate record from western Tibet. *Nature*, **353**, pp. 742–745.
1997   with Barker, P.A., Gell, P.A., Fritz, S.C. and Chalié, F., Diatom-inferred salinity in palaeolakes: an indirect tracer of climatic change. *Quaternary Science Reviews*, **16**, pp. 547–563.
1998   with Van Campo, E., A 40,000–yr pollen and diatom record from Lake Tritrivakely, Madagascar, in the southern tropics. *Quaternary Research*, **49**, pp. 299, 311.
2000   Hydrological changes in the African tropics since the Last Glacial Maximum. *Quaternary Science Reviews*, **19**, pp. 189–211.
2001   Hydrological changes in Africa. *Science*, **292**, pp. 2259–2260.
2004   with Roberts, N., Late Quaternary hydrologic changes in the arid and semi-arid belt of northern Africa. Implications for past atmospheric circulation. In *The Hadley Circulation: Present, Past and Future*, edited by Diaz, H.F. and Bradley, R.S., (Dordrecht: Kluwer Ac. Pub.), pp. 313–345.
2011   with Vidal, L., Develle, A.-L. and Van Campo, E., Hydrological variability in the Northern Levant: a 250 ka multiproxy record from the Yammoûneh (Lebanon) sedimentary sequence. *Clim. Past*, **7**, pp. 1261–1284. doi:10.5194/cp-7-1261-2011.
2014   Reminiscences and acknowledgements from a lover of deserts near the end of her professional life. *J. Paleolimnology*, **51**, pp. 139–144.

1.2.2 References about Françoise Gasse

EGU (European Geophysical Union), 2010, *Françoise Gasse, Hans Oeschger Medal 2010*. http://www.egu.eu/awards-medals/hans-oeschger/2010/francoise-gasse/
Sylvestre, F. and Batterby, R., 2015, *Key-note on the achievements of Françoise Gasse*, 1 p., http://www.insu.cnrs.fr/node/4827, and http://paleolim.org/index.php/2014/04/24/words-from-dr-florence-sylvestre-on-the-passing-of-francoise-gasse/

CHAPTER 2

Exploration of the Tibesti Mountains– Re-appraisal after 50 years?

Baldur Gabriel

Berlin, Germany

ABSTRACT: The Tibesti Massif is the highest and largest mountain area in the Sahara. With more than 3400 m a.s.l., it must have reached into different climatic zones during cold periods of the Quaternary. Hence, for the whole of Northern Africa, it acted as an ecological niche and as a refuge during the arid phases; while during the humid "pluvials", it operated as a centre for propagation. The former investigations into palaeoclimate, landscape, and cultural history, carried out mostly by French scientists and by the German Research Station at Bardai (established by the Free University of Berlin), came to an end after 1970 due to political problems. Only recently, researchers from the University of Cologne have resumed geoscientific research in this large, to date little-known area. This paper explains some of the previous activities and results and highlights important gaps in present knowledge. The history of climate and landscape evolution can better and more precisely be understood, in particular, by additional investigations into fluvial terraces (which document successions of erosion and accumulation) as well as into processes of weathering (fossil soils and their stratigraphic interference and intercalation by volcanic activities), and into the analysis of fossil-bearing sediments of ancient lakes, especially in high-altitude volcanic craters. In respect to cultural history the study of archaeological remains can help to solve hitherto unsettled questions: Where did people, animals, and plants, which populated the Sahara after the beginning of the Holocene humidification, come from? Where was the centre of the early African ceramic production? Where can we localize the origin of the wide-spread African cattle-herders? – There have long-since been serious suppositions and theories that the Tibesti was the main source of the beginning and further evolution of the Neolithic in Northern Africa. By this it must have fundamentally influenced the civilizations and state societies of the Old World that later evolved.

2.1 INTRODUCTION

German research in the North African dry belt has quite a long tradition. Already by the end of the 18th century Friedrich Hornemann crossed the Sahara, travelling from Cairo via Siwa and Murzuk up to the Niger River (Hornemann, 1802), followed during the 19th century by lengthy voyages of Heinrich Barth (Barth, 1857–58), Gustav Nachtigal (Nachtigal, 1879–89), and Gerhard Rohlfs (Rohlfs, 1868, 1881). Between World War I and World War II especially Leo Frobenius (Frobenius, 1963), together with Hans Rhotert (Rhotert, 1952), brought interesting results from their long expeditions. After World War II, in 1954–55, it was Wolfgang Meckelein from the Berlin Geographical Society, who managed an extended journey to the central Sahara, also to Wau en-Namus, and nearly up to the northern edge of the Tibesti Mountains (Meckelein, 1959) (Figure 1).

At that time, the Tibesti was rather inaccessible and accordingly rather disregarded by scientists, though it was with 3415 m a.s.l. the highest and with 150,000 km² the largest mountain range in the centre of the Sahara (Gabriel, 1991). Since 1899, it was part of the French colonial administrative authority. In fact it was not occupied before 1914, but gradually this was followed by its exploration (Gabriel, 1973). In 1921 already, Jean Tilho, a geographer from the French military service, published a detailed topographic map of the massif (Tilho, 1921). The geology, in general, was investigated first by Marius Dalloni (Dalloni, 1934–35) and later on – predominantly the volcanic phases – by Pierre Vincent (Vincent, 1963) and others. Extensive botanical research activities were undertaken by Pierre Quézel (Quézel, 1958). Charles LeCoeur explored the language, the habits, and the ethnographical situation of the people (LeCoeur, 1950, 1956; cf. Chapelle, 1957). However, the evolution of climate, landscape, and civilizations during the Quaternary was just mentioned as side-issues in marginal notes or – mainly various prehistoric data and discussions – communicated by enthusiastic members of the military staff like Paul Huard (Beck and Huard, 1969; Huard, 1978).

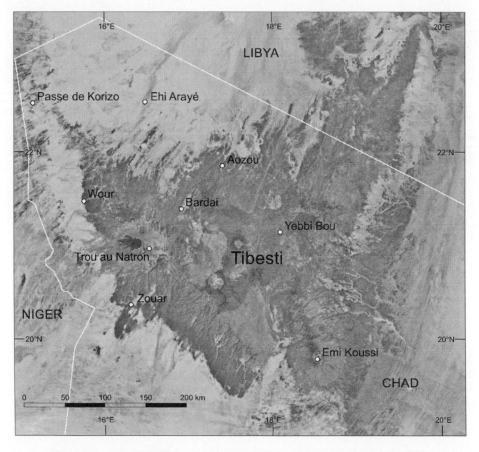

Figure 1. Annotated satellite map of the Tibesti Mountains in Chad – Google Earth; Images © 2017 Landsat/Copernicus.

2.2 THE RESEARCH STATION OF THE FREE UNIVERSITY OF BERLIN IN BARDAI

Therefore, in 1964, the building of a research station by Jürgen Hövermann in Bardai, in the administrative centre of the Tibesti, was quite important and meritorious (Hövermann, 1965). He was professor of geography and the head of the Lab for Geomorphology at the Free University of Berlin. Hövermann had previously investigated the glacial evolution of climate and landscape in central Europe as well as in the high mountains of Ethiopia and he was particularly interested in the effects of the adiabatic gradient: Air temperature depends on the height above sea level, the higher – the cooler (approx. 1°C/100 m), and because cold air cannot hold water vapour to the same degree precipitation must increase. Above the freezing point they fall as snow which can aggregate to form snow patches and glaciers and these – gliding downhill – can form trough valleys and nivation hollows like cirques, leaving different kinds of moraine deposits. Additional hints on former climate can be soil patterns from seasonal frozen ground (Gelisol) as in arctic zones or a change of climate can also be ascertained from special distinguishable marks of weathering, erosion, and sedimentation.

As a high mountain area situated in the subtropical dry belt, the Tibesti must have participated in several climatic zones throughout the Quaternary, and climatic changes should be traceable quite easily. Therefore, the massif promised a great variety of palaeoclimatic indicators as well as of ecological niches. Besides it was scarcely explored and it was situated within the reach of central Europe. Such considerations influenced the decision to establish the "Außenstelle des Geomorphologischen

Figure 2. The building of the "Maison allemande" on the southern outskirts of Bardai, surrounded by a fence, seen from the east. The main entrance was from the right, from the centre of the oasis. Over the left corner of the roof, the water reservoir, in white, of the local well is to be seen. The separate small building to the left was the kitchen. – Photo: B. Gabriel, 23.02.1967.

Figure 3. During the journey from Tripoli to Bardai: Jürgen Hövermann (left) and Karlheinz Kaiser use lunch hour for chess game in the shadow of a tamarisk tree. – Photo: B. Gabriel, Sept. 1966.

Laboratoriums der Freien Universität Berlin" (in English: the outpost of the geomorphological laboratory of the Free University of Berlin) – as it was called officially – in Bardai, in the heart of the mountain area at an altitude of 1018 m a.s.l.

From the beginning, the focus was laid on investigations concerning the relations of the Quaternary ice ages of Europe and the pluvial episodes in the Sahara (Hagedorn, 1982; Jäkel, 1977). No doubt, many prehistoric finds, fossil bones, and plant relics had long since proven that there had been wet phases, during which the Sahara was full of life and by no means such a hostile dry desert as today. However, controversial discussions resulted from exact dating or from the reasons and dimensions of such alterations in climate and landscape (Gabriel, 1982). Perhaps the deterioration (desertification) of these landscapes and conditions for life was reversible? Then, the research program would even gain an aspect of economic aid to developing countries that struggle for future welfare. Nevertheless, the first and main objective of the station was purely research.

The field work in the mountain area produced many scientific results. According to the main objective, most of the publications brought into focus problems of palaeoclimate and geomorphology. Up to nine consecutive phases of erosion and accumulation were distinguished within the large fluvial systems, in the "enneris" (wadis) Bardagué, Yebbigué and Misky. Some could be correlated to volcanic eruptions, to slope development (e.g. pediments), to soil formation, to prehistoric sites, to radiocarbon dates, to relics of fossil fauna and flora, or to palynological results. The outcome of fieldwork in geodesy, geology, prehistory, human geography, botany or meteorology were equally published (Jäkel, 1982). Since 1964, a weather-station recorded all data on weather conditions at 1020 m a.s.l. To register the climatic conditions in higher altitudes, a small data printer was installed at 2400 m a.s.l. at the edge of the crater of "Trou au Natron". It worked from April 1965 till August 1968 and showed, for example, a ten times higher rainfall than in Bardai (Heckendorff, 1977). Unfortunately, these instruments were stolen, later on the weather station was permanently supervised by a guard.

Every six months, the station crew was exchanged, leaving one member behind. Normally they were 4–5 graduate students, one technician, and the group leader, who was an experienced geo-scientist who coordinated and supervised field work. The new team came from Berlin via Italy by ship to Tripoli in Libya, where the returning crew handed over the vehicles. The supply for the next half year (foodstuff, fuel, technical equipment, etc.) had to be organized and stowed in the cars, if required also in an additionally hired lorry.

It took several days to reach Bardai, travelling either the official route via Sebha in Fezzan to the Pass of Korizo, the Enneri Tao, and the Tarso Toussidé ("Trou au Natron"), finally passing Oudingueur and Gonoa. Sometimes they departed from Sebha and used a turn-off across the Serir Tibesti to the landmark Ehi Arayé and up the river bed of Enneri Bardagué. The journey was always an exciting and important introduction into the problems of deserts for the newcomers.

However, due to political trouble in the northern part of Chad since 1968, these possibilities of approach were cut off. Now Bardai was only reachable by plane from Fort Lamy (today N`Djamena) via Faya Largeau. Field work became unsafe, so the research program as well as the number of team members was reduced. After 1973, especially as a consequence of a spectacular kidnapping in Bardai in April 1974 (cf. Desjardins, 1975), the station was totally abandoned. Until then more than 50 team members had done intensive field work, and the "Mission allemande" (Figure 2) – as it was called in Tibesti – had been a hospitable place to stay for many international scientists and also for some tourists.

Through people, ideas and experience, the research project influenced many of the later academic teams in Germany to deal with arid areas: Since the early 70ies, the "Africa group" in Würzburg around Horst Hagedorn, who had been the first station head in Bardai 1964 (Hagedorn and Busche, 1982; Hagedorn and Wagner, 1979); since 1974, the "Desert Working Group" in Stuttgart, assembled by Wolfgang Meckelein (Meckelein, 1977); after an impressive international Sahara-exposition in 1978 by the Rautenstrauch-Joest-Museum in Cologne, the team B.O.S. (Besiedlungsgeschichte der Ost-Sahara), organized by the prehistorian Rudolph Kuper (Kuper, 1978, 1995); and finally, since 1980 in Berlin, the Special Research Unit 69 on Arid Arcas of the German Research Foundation DFG (SFB 69 der Deutschen Forschungsgemeinschaft), initiated by the geologist Eberhard Klitzsch (Gabriel, 2001; Klitzsch, 1980, 2004).

Repeated attempts at re-organizing the work in Bardai failed. However, just recently, in 2015, Stefan Kröpelin, a geographer from Cologne University, who had already been a member of the Berlin SFB-team and of B.O.S. (Kröpelin, 1993, 2007), succeeded in leading a geo-scientific expedition to the Tibesti Mountains. Starting from Bardai, they were able to investigate and collect sediment samples for palaeoclimatic data in the volcanic craters of the "Trou au Natron" and of Emi Koussi (Kröpelin 2015). A second expedition in February-March 2016 supplemented this encouraging new start. Kröpelin was generously supported by the Government of Chad, who revealed strong interest in a revival of scientific investigations in Tibesti.

2.3 THE PALAEOECOLOGICAL SIGNIFICANCE OF THE MOUNTAIN MASSIF

Indeed, a revision of the previous work would be a great progress and would promise an enormous profit for geo-scientific research. Many problems remained unsolved or have turned up later on in retrospect. Besides, over the last 50 years, the scientific world has advanced in knowledge and methods. Finally, we are now pushed to suppose that the key for the solution of several questions in pre-history are to be found in Tibesti. Important radical changes and innovations for Old World antiquity must have had their sources in this mountain massif.

Large unstructured plains like the serirs or regs in the Sahara are quite homogenous in ecological respect: Air masses of any quality can move without hindrance, the irradiation is the same everywhere, mass movements at the surface are confined, they have only a limited range of space or quantity, and just different soil quality can occasionally change the vegetation cover. Therefore, an economic benefit, an exploitation by animals or humans, is impossible if the regional or zonal climate is as unfavourable as today (Gabriel, 1984).

In mountain areas many special locations in micro-, meso- and macro-scale, each with prevailing ecological conditions, result in a great variety of species, habitats, and chance of survival: caves, rocky overhangs ("abris"), windward or leeward locations, shadow and sunshine slopes, adiabatic change of temperature and precipitations, together with variable geologic, soil surface and ground incline circumstances. The mountain climate facilitates an easier survival of humans and animals by seasonal migration (transhumance). Springs, valleys, pools, ponds, or lakes supply people and animals effortlessly with water; and in case of social conflicts, the hostile groups can make use of small scale topographical situations for hiding or for the purpose of defence.

Indigenous people often see prominent landscape elements in connection with myths and as traditional symbols: mountain peaks, volcanoes, hot springs, waterfalls or other special locations and conspicuous landscape elements are considered to be the residence of gods and ghosts. They can be places of pilgrimage and assembly, or they can contribute to medical curing of illnesses, like the hot springs of Soborom in Tibesti.

Geologic outcrops on slopes and ridges or elsewhere can easily provide stone material and mineral resources, which is especially important for Stone Age technologies and civilizations. On the great sandy plains of the central desert, on the contrary, the Neolithic cattle herders had to transport their heavy tethering stones (of more than 10 kg) over large distances (cf. Close, 1997). There, raw material for smaller artefacts was sometimes found in gravel deposits, but again the components of their typical fire places, the "stone sites" (Steinplätze) were scarce, as well as superstructures of tombs.

Consequently, the highest and most extended massif in the centre of the largest desert area on earth must have reacted adjusting to the variations of climate. Always during extreme arid phases, it must have retreated to the mountains, from where the plants, animals and humans could spread out again, if during more humid periods ("pluvials") their existence was no longer endangered in lower hilly countries or plains. For the last 30,000 years, i.e. for the Late Pleistocene and the Holocene, several such climatic reversals are established not only for the Ice Age areas of central Europe, but also for the subtropical arid belt. It is the same time during which the *Homo sapiens* co-developed and progressed from a multiform Late Palaeolithic to the Neolithic Revolution and to the advanced civilizations of antiquity.

Such considerations lead to realize that the Tibesti Mountains must have been a preserve of relics during arid phases and a source of new life in the subtropical belt during the pluvial episodes.

2.4 PROBLEMS OF CULTURAL HISTORY

Nobody knows where the people who populated the Sahara at the end of a long dry period at about 12,000 B.P. came from. Did they migrate along with the monsoon rain frontier from the south, from the interior of Africa or did they move in from the north (Mediterranean)? Maybe they even arrived from the east, from the Near East or from the Nile valley, where they could have survived and persisted during arid phases? Several indications point to the fact that Black African as well as Mediterranean people participated in early human societies, but they might have developed different survival strategies, and might possibly have mixed with each other when they lived together in the higher mountain areas during arid periods (Gabriel, 1987). The present population of Tubu and Goranes must be the descendants of them.

Similarly unknown is the origin of huge cattle herds that traversed the interior plains of the Sahara – green like savannas during the Neolithic wet phase (Gabriel, 2002a, 2002b; Jordeczka *et al.*, 2013; Kuper and Riemer, 2013). Until recently it was supposed that the process of domesticating animals had happened in the "Fertile Crescent" and, therefore, the horned cattle must have been introduced from Near East. Now it seems clear that the African cattle – except for the Zebu race – are of local stock. Where did they find enough water and feed to survive during the arid inter-pluvials? Perhaps in the Nile valley which, however, was deeply and extensively flooded during several months of the year? Maybe in the Ethiopian Highlands or in the Atlas Mountains, respectively at the Mediterranean coastal zone? Did they originate from inside of Africa, where indeed the tsetse fly would have limited their regional spreading and distribution in the region?

In any case the Tibesti granted a lot of favourable ecological niches like the tarsos, which are large plains at more than 2000 m a.s.l.; also fertile plains and terraces along the big enneris; and many intramontaneous sandy alluvial plains (Sandschwemm-ebenen). Additional arguments and considerations permit, therefore, the conclusion that the nomadic cattle herders who are still active and wide-spread all over Africa have had their origin within the mountain areas of the central Sahara at about 10,000 B.P.

Figure 4. On the occasion of the national public holiday, different military units, school children, and young women and girls' league dressed in the colours of the national flag marched through the centre of Bardai under the watching eyes of the older local notabilities. – Photo: B. Gabriel, 11.01.1967.

Figure 5. After numerous rock art evidence – like here from Bardai – domestic cattle played an important role in the life style of Neolithic societies not only in the Saharan great plains, but also in the mountains. However, exact chronological data are still missing. – Photo: B. Gabriel, Nov. 1966.

The process of domestication was obviously independent of parallels in the Near East (Gabriel, 2002a).

Where was the (African) beginning of pottery production c. 10,000–12,000 years ago (cf. Close, 1995; Gabriel, 1981; Jesse, 1999; Jordeczka *et al.*, 2011)? Of course the invention was not earlier than in the Near East or in the Far East, but we can be sure it occurred independently! On the one hand, the earliest radiocarbon dates originate from the Western Desert of Egypt and from the Sudanese Nile area; on the other hand, from southern Algeria and northern Niger. These two different domains are separated by an apparent hiatus which is occupied by the huge, barely investigated area of the Tibesti Mountains.

The only site where more than 8000 years old pottery was found in Tibesti is Gabrong, some 30 km to the northeast of Bardai (Gabriel, 1977, 1978, 1981). Here, the cultural layer below lake sediments contained a bowl which strikingly showed both common decoration styles of the ancient pottery – the "dotted wavy line" pattern covered the upper part of the vessel and the "rocker stamp" (made by a tooth stick) the lower part. The latter Neolithic decoration type is known from all over the North African dry belt, from the Atlantic to the Red Sea. Charcoal originating from the Gabrong find layer yielded a radiocarbon date of about 1000 years later than those in the adjacent areas, but a second posterior sample from the same site was only 500 years later. Is it daring or wrong to suppose that the deficiency of data is just a gap in research? And that the Tibesti in reality had been the centre, from where the new technique spread out to East and to West?

During the last decades, it became more and more obvious that the North African prehistoric populations had influenced the later advanced civilizations in the Nile valley and in the Mediterranean more than was known up to now (cf. Kuper and Keuthmann, 1991). At times, the cultural trend was directed west-east and south-north

Figure 6. The Neolithic site of Gabrong: Beneath a rocky overhang, a profile with lake sediments had been preserved (left of the image centre). They are horizontally stratified and passed through by fossil rhizomes of reed. Below them existed a cultural find layer with stone artefacts, a shoulder blade of a buffalo, and fragments of a pottery bowl. – Photo: B. Gabriel, 13.10.1966.

rather than inverse. Even linguistic details may realize that still in Roman antiquity, the Mediterranean world kept behaviour patterns and attitudes of mind from Africa. Since Neolithic times until today, the identical root for the denotation of cattle (*pecus*) and for money (*pecunia*) has the same meaning in African pastoral tribes.

The antique territory of the Garamantes in Fezzan with their municipal centre of Germa must have incorporated the Tibesti area, too. Large graveyards of their characteristic "chouchet"-tombs near Zoui, Aozou, and elsewhere give evidence of a large population at that time, though their lifestyle, their housing patterns or their economic base are still unknown (cf. Rönneseth, 1982:51). What kind of relations did they have with a supposed central government in Fezzan? In Tibesti, at least, ruins of an antique urban centre like Germa, which could have pointed to an autonomous higher social organization, are unknown. Nevertheless, some information derives from contemporary rock art. Many depictions represent armed camel riders.

Actually the knowledge about the different burial practices and the variety of grave types in Tibesti during the past millennia is still rudimentary (Gabriel, 1970, 1999), although such practices can offer a great deal of information about prehistoric civilizations. They can supply many valuable data with respect to medicine, demography or physical anthropology; to understanding of religious conceptions and ideas; to technical abilities and skill; to economic facility or pressure; to social factors of organization and kinship; to supra-regional connections and exchange; to progress and rate of alterations in societies without any written tradition; and to changes in their natural

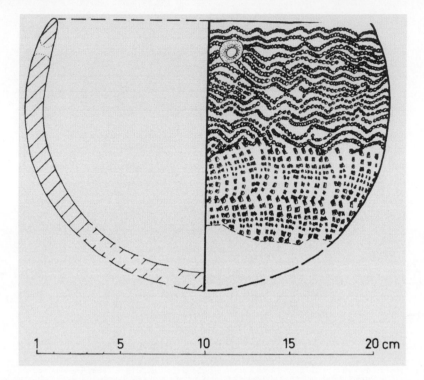

Figure 7. The more than 8000 years old bowl from Gabrong is decorated with the most common ornamentation styles of the early pottery production. The upper part is covered with a "dotted wavy line" pattern and the lower part with horizontal bands of "rocker stamp" impressions, made by a tooth stick ("Zahnstockmuster in Wiegebandtechnik"). A perforation of the rim of the vessel was presumably used for fastening a string. – Drawing: B. Gabriel.

Figure 8. The huge graveyard of the Garamantes, wide-spread over hills and valleys north of Zoui, has an extension of more than 1 km. The typical "chouchet"-tombs are round towers of varying height and diameter. Their vertical or conical walls are built from slabs, boulders, and blocks; and the interior is filled with smaller material like gravel, debris, and sand. Sometimes the tombs can contain burials of several individuals. – Google Earth, Images © 2017 CNES/Airbus, Centre of the Image: 21°20′49″N, 17°05′17″E.

Figure 9. Rock art panel near Gabrong, Enneri Dirennao: If the representations of the large cattle herd and the small camel below them are contemporary (which seems probable from the degree of patination), cattle-keeping during the time of the Garamantes must have been still significant. The somewhat schematic style is in distinct contrast to the obviously earlier (Neolithic?) image of Figure 5. – Photo: D. Busche, Nov. 1966.

Figure 10. A pre-islamic "ring grave" (No. 10) from a small cementery near Gonoa. A circular rampart surrounded a pear-shaped hole, which had been carefully carved into the soft ignimbrite and in which the deceased was buried in a sitting position. The hole was covered by trimmed ignimbrite slabs and surrounded by a ring of ignimbrite fragments, nicely formed to crescents, sometimes also by upright standing slabs. – Photo: B. Gabriel, 01.12.1966.

environment. Together with investigations into the technological and stylistic sequence of ancient pottery as well as the analysis of contents and style of the manifold rock art, such inquiries would result in a chronological frame of the numerous civilizations in the Tibesti Mountains and their surroundings. For instance, the relations of the "ring graves" of Gonoa with the famous Neolithic rock art site nearby could easily secure their synchronism (Gabriel, 1970; Staewen and Striedter, 1987).

Accordingly archaeologists and prehistorians would find a great and interesting scope of activity. For instance, a systematic sequence of the different stone artefacts is equally still missing. Palaeolithic tool kits from flake and core industries occur as wide-spread surface finds in the mountains (Dalloni, 1934–35; Tillet and Gabriel, 1990). Nevertheless, their precarious typological chronology should be corroborated by stratigraphic data, by a sequence of layers, for example, in caves or below volcanic deposits, with facilities for radiometric analysis.

Moreover, what about the obsidian, which is very common in Tibesti? Did Palaeolithic people already use it? Until when after the Neolithic period was this material exploited in mines (Gabriel, 1979)? To where was it exported? Owing to its hardness and its good splitting facility, the volcanic glass is known as one of the earliest trading commodity of mankind. Other minerals, like the green amazonite for antique pearls, are supposed to have had their origin in Tibesti (cf. Monod, 1974).

Already more than 60 years ago, A. J. Arkell concluded, "It seems, therefore, in the present state of our knowledge that the Tibesti area may have been the centre from which the Fayum, Shaheinab and Ténéré cultures were diffused ... such a hypothesis will at least justify a detailed archaeological survey of the Tibesti area ... the result will be ... important to the study of the rise of the Neolithic ..." (Arkell, 1955: 346).

2.5 PROBLEMS OF CLIMATE AND LANDSCAPE HISTORY

Apart from questions in cultural history, there are many challenges dealing with the development of climate and landscape. Not very much is known about the chronology of the volcanic events, beginning in Cretaceous and obviously still active during the Holocene (Vincent, 1963). Especially the successions of basaltic lava and ignimbrite eruptions, separated by fossil soils or as a result of erosion and sedimentation, should be thoroughly investigated. For most of the students and doctoral candidates who mapped and investigated special Enneris or parts of them, a follow-up examination of the fluvial terraces is important. They all found different systems of alternating periods with erosion and accumulation. Up to nine cycles were discerned, and a final concordance by comparing and checking them in the field is a significant desideratum.

The most interesting of these cycles is the so-called "middle terrace" ("Mittelterrasse", cf. Gabriel, 1972a, 1977; Jäkel, 1971; Jäkel and Schulz, 1972). It is the only one that is largely composed of limnetic sediments rich in carbonate, molluscs, and micro-fossils. Therefore, the ecology of these former lakes is well determined. But the difficulty is in explaining how in mountain-rivers with a certain gradient large stillwater bodies could evolve. What was the reason, the trigger (cf. Kaiser in Böttcher *et al.*, 1972)? A damming up by volcanic lava, by shifting sand dunes, by lateral influx of sediments from tributaries or by landslides from the slopes? Each of these would have accounted only for single events, but the phenomenon was observed throughout the Tibesti.

Finally, after comparing with similar situations in other areas, a hypothesis was offered, which could not be verified further in the field because of the political insecurity. In Plitvice in Croatia, Afghanistan, South Africa, California and other parts of the world, the course of rivers can be barred by sinter dams crossing the valleys.

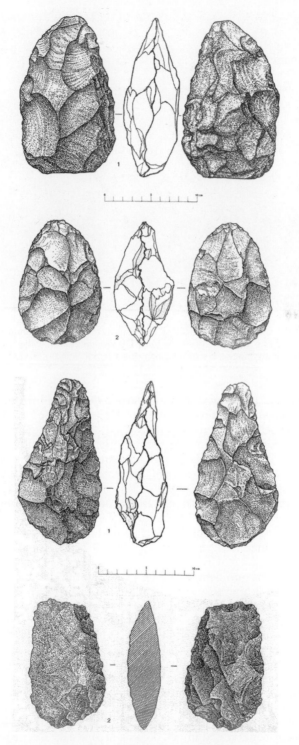

Figure 11. Three-hand axes and a cleaver from a Lower Palaeolithic surface site at Ehi Kournéi near Bardai. – Drawing: E. Hofstetter.

Figure 12. About 40 km northeast of Bardai four fluvial terraces of gravel accumulations accompany the valley of the Enneri Dirennao in different heights. The lake sediments of the "Mittelterrasse" are still missing at this locality, they occur only further downstream. - Photo: B. Gabriel, 24.06.1967.

Was it the same in Tibesti? At least, by radiocarbon dates and by finds of cultural relics within the accumulations at several sites, the "Mittelterrasse" could be attached to the Neolithic humid period (c. 12,000–6000 B.P.).

The process of formation and the exact chronological position of the "High terrace" (or the "Hohe Verschüttungen" – High blocking sedimentation) in the lower part of the Enneri Bardagué and in its affluent remained obscure (Obenauf, 1972). The accumulations consist of more than 20 m stratified gravel, sand, and silt with a high proportion of volcanic pyroclastic material like pumice, cinder, and ashes. Certain indications show parallels to the Mittelterrasse, for instance calcareous layers with fossils, but it seems to be doubtful if still during the early Holocene such a strong volcanic activity happened in the surrounding district of Pic Toussidé, which blocked the discharge.

Another precarious problem is the existence of thick limnetic accumulations within volcanic craters at high altitudes like at the "Trou au Natron" or at Emi Koussi. Obviously the former lake level within the "Trou au Natron" was more than 400 m higher (Böttcher *et al.*, 1972). Does it mean that these lakes developed by precipitation only? The enormous depth of this lake suggests that the lake level lowered from top to bottom without having any lateral influx from the summit region (i.e. an exclusively rainfed bassin?). Annual rainfall must have been extraordinarily high and evaporation must have been strongly reduced because of lower temperatures!

An explanation could be that the original depth of the lake was quite normal – maybe up to several tens of meters – and only afterwards the bottom sank down by a phreatic explosion to nearly a depth of 1000 m; while up at the walls of the caldera, some of the lake sediments rested intact. Then a second lake developed further down and left a younger series of accumulations. Early radiocarbon dates support such a

Figure 13. The accumulations of the "Mittelterrasse" are composed of lake sediment layers with varying ecological evidence: Here the calcified roots of reed at the upper part of a profile in the Enneri Dirennao document a shallow water level. – Photo: B. Gabriel, 06.05.1967.

view, for – according to them – the upper series is about 2400 years older than the lower one. Within this space of time during the early Holocene, the sinking of the bottom may well have happened.

Even if this lowering process is not explained by an explosion, but by an implosion, a multiphase formation of this huge depression without any outlet seems to be an expedient explanation to understand the different altitudes and ages of the lake sediments. The only natural forces – other than from endogenous tectonic energy – to counteract gravity and to remove solid rock and sediments out of closed depressions, are wind and water eddies, ice pressure or organisms, but never in such dimensions. The evolution of the "Trou" should, therefore, have experienced at least three stages of deepening, each interrupted by lake sedimentation of several metres. Probably Stefan Kröpelin and his team will bring some light to this problem.

Research in flora and fauna of the Tibesti can be widely improved, even though the knowledge of the stock of plant species and of vegetation habitats and units is fairly advanced, mainly contributed to by French botanists (e.g. Maire and Monod, 1950; Quézel, 1958). Nevertheless, special locations may still offer novelties, for instance, many narrow ravines deeply incised in the ignimbrite of the Tarso Toussidé or small moist localities and water points in the eastern part of the Tibesti, which is rather inaccessible and less explored. Moya (several km north of Aozou) is an example of such a special situation, where water is leaking out of a vertical sandstone wall covered by a curtain of moss, fern, and other plants.

Attention should be laid on former and present fauna. Not only by rock art, but also by fossil finds, it is evident that elephants and giraffes were still living in the Tibesti during the early Holocene. Even relics of Pliocene mastodons were found (Gabriel, 1972b), but the sample of bones was collected only superficially at the site

Figure 14. The "Trou au Natron" is a nearly 1000 m deep volcanic crater at the foot of Pic Toussidé. The bottom is covered by salt marsh, from which (at the left) a secondary volcanic cone rises up to 100 m. The persons who are coming uphill are (from left to right): Detlef Busche, Jürgen Hövermann, Frieder von Sass, Jörg Grunert and – on the right side – Karlheinz Kaiser. – Photo: B. Gabriel, 08.10.1966.

near Puits Tirenno (between Bardai and Aozou), and therefore an intensive and careful investigation seems to be worthwhile.

2.6 FINAL REMARKS

Of course, the interest of the Government of Chad and the local authorities of Bardai in reorganizing scientific research in Tibesti is directed towards economic aspects and application. Nevertheless, basic work in meteorology and geology, in prehistory and archaeology, and in other scientific branches should be completed.

Mineral resources, for example, seem still to be awaiting exploration and exploitation. After his investigations, Stefan Kröpelin observed a gold rush in the area of Enneri Misky (Kröpelin, 2015). The deposits of uranium and tungsten in Yedri hills (north of Aozou; Vincent, 1969) have been known since long, but they seem to be not very rich and difficult to be exploited. In the meantime uranium has lost some of its global importance in the last 60 years. What about the widespread occurrence of ignimbrite? Like volcanic tuff or pumice stone in Italy or in the German Eifel area, they could serve as building material if need arises in the future and if an appropriate branch of industry and transport means develop.

Indeed, projects for economic development are important for the local population. The infrastructures of traffic and transport, of communication, and of administration have to be improved as well as the supply of medical, educational or marketing institutions. Tourism would bring additional income and stimulation. Deficiency of water has always been a serious problem. In similar arid areas, small dams in wadis have facilitated a 'modus vivendi' for the people. This could easily be arranged in the Tibesti Mountains, too. Anyhow, the amount of rainfall in the high mountain zones has been gauged at about 100 mm/a. episodically, and in sections, devastating floods can therefore pour through the valleys (Figure 15).

Figure 15. A dam was built in Wadi Abu Dom (Bayuda, northern Sudan) near Bir Merwa in 2013, which after only two years led to a fresh water lake of more than 2 km². – Photo: B. Gabriel, 28.02.2015.

But it may take some time for a realization of such ideas, more so because the situation has not yet become fully stabilized. War relics are mine fields, destroyed villages (like Yebbi Bou), deserted palm groves, and abandoned horticulture. A prerequisite is that the people experience and get accustomed again to fairly normal and safe conditions to gain confidence in their future life. Hopefully then, research in the Tibesti Mountains can prosper as it was before its collapse 50 years ago, to the benefit of the inhabitants and of the scientific world.

ACKNOWLEDGEMENT

Many thanks to Dr Gwyneth Edwards (Cardiff/Wales) for scrutiny and correction of the English version.

REFERENCES

Arkell, A.-J., 1955, The relations of the Nile Valley with the Southern Sahara, *IIe Congrès Panafricain de Préhistoire Alger 1952*, pp. 345–346.

Barth, H., 1857–1858, *Reisen und Entdeckungen in Nord- und Central-Afrika in den Jahren 1849–1855*, Gotha: Justus Perthes, 5 vols.

Beck, P. and Huard, P., 1969, *Tibesti – Carrefour de la préhistoire saharienne*, Paris: Arthaud, pp. 1–292.

Böttcher, U., Ergenzinger, P.-J., Jaeckel, S.H. and Kaiser, K., 1972, Quartäre Seebildungen und ihre Mollusken-Inhalte im Tibesti-Gebirge und seinen Rahmenbereichen der zentralen Ostsahara, *Zeitschrift für Geomorphologie N.F.* (Berlin-Stuttgart), **18**, pp. 182–234.

Chapelle, J., 1957, *Nomades noirs du Sahara*, Recherches et Sciences Humaines (Paris) **10**, pp. 1–449.

Close, A.E., 1995, Few and Far Between – Early Ceramics in North Africa, *The Emergence of Pottery* (Washington and London), pp. 23–37.

Close, A.E., 1997, Lithic economy in the absence of stone, *Journal of Middle Atlantic Archaeology,* **13**, pp. 27–56.

Dalloni, M., 1934–1935, *Mission au Tibesti (1930–1931)*, Mémoires de l'Académie des Sciences de l'Institut de France (Paris), **61/62**, 2 vols.

Desjardins, T., 1975, *Avec les otages du Tchad*, (Paris: Presses de la Cité), pp. 1–288.

Frobenius, L., 1963, *Ekade Ektab, Die Felsbilder Fezzans,* (Graz: Akademische Druck- und Verlagsanstalt), pp. 1–74 + 91 Tafeln.

Gabriel, B., 1970, Bauelemente präislamischer Gräbertypen im Tibestigebirge (Zentrale Ostsahara), *Acta Praehistorica et Archaeologica* (Berlin), **1**, pp. 1–28.

Gabriel, B., 1972a, Terrassenentwicklung und vorgeschichtliche Umweltbedingungen im Enneri Dirennao (Tibesti, östliche Zentralsahara), *Zeitschrift für Geomorphologie N.F.* (Berlin/Stuttgart) *Supplement*, **15**, pp. 113–128.

Gabriel, B., 1972b, Zur Vorzeitfauna des Tibestigebirges, *Palaeoecology of Africa*, **6**, pp. 161–162.

Gabriel, B., 1973, Von der Routenaufnahme zum Weltraumphoto. Die Erforschung des Tibesti-Gebirges in der Zentralen Sahara, Berlin, *Kartographische Miniaturen*, **4**, pp. 1–96.

Gabriel, B., 1977, Zum ökologischen Wandel im Neolithikum der östlichen Zentralsahara, *Berliner Geographische Abhandlungen*, **27**, pp. 1–111.

Gabriel, B., 1978, Gabrong, Achttausendjährige Keramik im Tibesti–Gebirge, *Sahara – 10.000 Jahre zwischen Weide und Wüste*, (Köln: Museen der Stadt Köln), pp. 189–196.

Gabriel, B., 1979, Ur- und Frühgeschichte als Hilfswissenschaft der Geomorphologie im ariden Nordafrika, *Stuttgarter Geographische Studien,* **93**, pp. 135–148 (Festschrift für Wolfgang Meckelein).

Gabriel, B., 1981, Die östliche Zentralsahara im Holozän, Klima, Landschaft und Kulturen (mit besonderer Berücksichtigung der neolithischen Keramik), *Préhistoire Africaine. Mélanges offerts au Doyen Lionel Balout,* Paris, pp. 195–211.

Gabriel, B., 1982, Die Sahara im Quartär. Klima-, Landschafts- und Kulturentwicklung. *Geographische Rundschau,* **34**, pp. 261–268.

Gabriel, B., 1984, Great Plains and Mountain Areas as Habitats for the Neolithic Man in the Sahara, In *Origin and Early Development of Food–Producing Cultures in North-Eastern Africa,* Poznan, edited by Krzyzaniak, L. and Kobusiewicz, M., pp. 391–398.

Gabriel, B., 1987, Palaeoecological Evidence from Neolithic Fireplaces in the Sahara, *The African Archaeological Review,* **5**, pp. 93–103 (Papers in honour of J. Desmond Clark).

Gabriel, B., 1991, Gebirgsregionen der Ostsahara, *Revue de Géographie alpine,* **79**, pp. 101–116.

Gabriel, B., 1999, Enneri Tihai – eine vorgeschichtliche Grabanlage aus Südlibyen, *Beiträge zur Allgemeinen und Vergleichenden Archäologie,* **19**, pp. 129–150.

Gabriel, B., 2001, Geological and palaeoecological investigations by the Berlin Research Group "Arid Areas" in Northeastern Africa. *Palaeoecology of Africa,* **27**, pp. 305–316.

Gabriel, B., 2002a, Alter und Ursprung des Rinderhirten-Nomadismus in Afrika (mit einer Literaturübersicht). *Erdkunde,* **56**, pp. 385–400.

Gabriel, B., 2002b, Neolithic camp sites in the Sahara, Anticipation of future research. *Africa Praehistorica,* **14**, pp. 51–66 (Festschrift für Rudolph Kuper).

Gabriel, B., 2012, Tethering stones and stone sites (Steinplätze) at the Fourth Nile Cataract. *Africa Praehistorica,* **22**, pp. 83–90 (Proceedings of the Third International Conference of the Fourth Nile Cataract Cologne 2006).

Hagedorn, H., 1982, Die Forschungsstation Bardai – ihre wissenschaftlichen Voraussetzungen und Grundlagen, *Berliner Geographische Abhandlungen,* **32**, pp. 7–12.

Hagedorn, H. and Busche, D., eds., 1982, Festschrift für Jürgen Hövermann, *Würzburger Geographische Arbeiten,* **56**, pp. 1–187.

Hagedorn, H. and Wagner, H.-G., eds., 1979, Natur- und wirtschaftsgeographische Forschungen in Afrika, *Würzburger Geographische Arbeiten,* **49**, pp. 1–325.

Heckendorff, W.D., 1977, Untersuchungen zum Klima des Tibesti-Gebirges, *Berichte des Instituts für Meteorologie und Klimatologie der Technischen Universität Hannover,* **17**, pp. 1– 347.

Hornemann, F.K., 1802, repr. 1996, *Tagebuch seiner Reise von Cairo nach Murzuk,* Hildesheim: Veröffentlichungen des Landschaftsverbandes Hildesheim e.V., **8**, pp. 1–240.

Hövermann, J., 1965, Eine geomorphologische Forschungsstation in Bardai/Tibesti-Gebirge, *Zeitschrift für Geomorphologie N.F.* (Berlin/Stuttgart), **9**, p. 131.

Huard, P., 1978, Die Felsbilder des Tibesti-Gebirges, *Sahara - 10.000 Jahre zwischen Weide und Wüste,* (Köln: Museen der Stadt Köln), pp. 272–278.

Jäkel, D., 1971, Erosion und Akkumulation im Enneri Bardagué-Arayé des Tibesti-Gebirges (zentrale Sahara) während des Pleistozäns und Holozäns, *Berliner Geographische Abhandlungen,* **10**, pp. 1–55.

Jäkel, D., 1977, The work of the field station at Bardai in the Tibesti Mountains. *Geographical Journal* (London), **143**, pp. 61–72.

Jäkel, D., 1979, Run-off and fluvial formation processes in the Tibesti Mountains as indicators of climatic history in the Central Sahara during the Late Pleistocene and Holocene, *Palaeoecology of Africa,* **11**, pp. 13–44.

Jäkel, D., 1982, Verzeichnis der aus der Forschungsstation Bardai/Tibesti erschienenen großmaßstäbigen Karten, Aufsätze (A), Mitteilungen (M) und Monographien (Mo), *Berliner Geographische Abhandlungen*, **32**, pp. 167–176.

Jäkel, D. and Schulz, E., 1972, Spezielle Untersuchungen an der Mittelterrasse im Enneri Tabi, Tibesti-Gebirge, *Zeitschrift für Geomorphologie N.F.* (Berlin/Stuttgart) *Supplement*, **15**, pp. 129–143.

Jesse, F., 1999, Zur Wavy Line-Keramik in Nordafrika unter Berücksichtigung des Wadi Howar (Sudan) und dort des Fundplatzes Rahib 80/87. *Archäologische Informationen*, **22**, pp. 99–104.

Jordeczka, M., Królik, H., Masojć, M. and Schild, R., 2011, Early Holocene pottery in the Western Desert of Egypt: new data from Nabta Playa, *Antiquity*, **83**, pp. 99–115.

Jordeczka, M., Królik, H., Masojć, M. and Schild, R., 2013, Hunter Gatherer Cattle-Keepers of Early Neolithic El Adam Type from Nabta Playa: Latest Discoveries from Site E-06-1, *African Archaeological Review*, **30**, pp. 253–284.

Klitzsch, E., 1980, Konzeption des geplanten Berliner Sonderforschungsbereiches "Geowissenschaftliche Probleme arider Gebiete", *Berliner geowissenschaftliche Abhandlungen (A)*, **24**, pp. 31–38.

Klitzsch, E. H., 2004, From Bardai to SFB 69: The Tibesti Research Station and later geoscientific research in Northeast Africa, *Die Erde*, **135**, pp. 245–266.

Kröpelin, S., 1993, Zur Rekonstruktion der spätquartären Umwelt am Unteren Wadi Howar (Südöstliche Sahara/NW-Sudan), *Berliner Geographische Abhandlungen*, **54**, pp. 1–293.

Kröpelin, S., 2007, The Saharan lakes of Ounianga Serir – a unique hydrological system. *Africa Praehistorica*, **21**, pp. 54–55.

Kröpelin, S., 2015, *Bericht über eine Forschungsreise ins Tibesti-Gebirge im Frühjahr 2015*. - Abendvortrag in der Gesellschaft für Erdkunde zu Berlin am 17-09-2015 (see also *Bild der Wissenschaft*, **9**, pp. 70–74.)

Kuper, R., ed. 1978, *Sahara – 10.000 Jahre zwischen Weide und Wüste*, (Köln: Museen der Stadt Köln), pp. 1–470.

Kuper, R., 1995, Prehistoric Research in the Southern Libyan Desert, A brief account and some conclusions of the B.O.S. project, *Cahiers de Recherches de l'Institut de Papyrologie et d'Ègyptologie de Lille*, **17**, pp. 123–140.

Kuper, R. and Keuthmann, K., 1991, Die Rolle der Kultur Afrikas in der Begegnung mit Europa, *Afrika und Europa*, Bochum: Interdisziplinärer Arbeitskreis für Entwicklungsländer-Forschung, Texte **2**, pp. 64–101.

Kuper, R. and Riemer, H., 2013, Herders before Pastoralism: Prehistoric Prelude in the Eastern Sahara. In *Pastoralism in Africa. Past, Present and Future*, edited by Bollig, M., Schnegg, M. and Wotzka, H.-P., (New York and Oxford: Berghahn Books), pp. 31–65.

LeCoeur, C., 1950, *Dictionnaire ethnographique téda. Précédé d'un lexique français-téda*, (Paris: Mémoires de l'Institut Français d'Afrique Noire), Dakar, **9**, pp. 1–211.

LeCoeur, C. and LeCoeur, M., 1956, *Grammaire et texts téda-daza*. (Paris: Mémoires de l'Institut Français d'Afrique Noire), Dakar, **46**, pp. 1–394.

Maire, R. and Monod, T., 1950, *Études sur la flore et la végétation du Tibesti*. (Paris: Mémoires de l'Institut Français d'Afrique Noire), Dakar, **8**, pp. 1–140.

Meckelein, W., 1959, *Forschungen in der zentralen Sahara. I. Klimageomorphologie.* (Braunschweig: Westermann), pp. 1–181.

Meckelein, W., ed., 1977, Geographische Untersuchungen am Nordrand der tunesischen Sahara. Wissenschaftliche Ergebnisse der Arbeitsexkursion 1975 des Geographischen Instituts der Universität Stuttgart, *Stuttgarter Geographische Studien*, **91**, pp. 1–300.

Monod, T., 1974, Le mythe de "l'émeraude des Garamantes". *Antiquités Africaines* (Paris), **8**, pp. 51–66.

Nachtigal, G., 1879–89, repr. 1967, *Sahara und Sudan*. Berlin, 3 vols. (repr. with preface, index etc.: Graz 1967).

Obenauf, K.P., 1972, Die Enneris Gonoa, Toudoufou, Oudingueur und Nemagayesko im nordwestlichen Tibesti. Beobachtungen zu Formen und zur Formung in den Tälern eines ariden Gebirges, *Berliner Geographische Abhandlungen*, **12**, pp. 1–70.

Quézel, P., 1958, *Mission botanique au Tibesti*, Mémoires de l'Institut de Recherches Sahariennes (Alger), **4**, pp. 1–375.

Rhotert, H., 1952, *Libysche Felsbilder*, (Darmstadt: Veröffentlichungen des Frobenius-Instituts der Universität Frankfurt/M.), pp. 1–146.

Rohlfs, G., 1868, *Reise durch Nordafrika vom Mittelländischen Meere bis zum Busen von Guinea 1865–1867. 1. Hälfte: Von Tripoli nach Kuka (Fezzan, Sahara, Bornu)*. Petermanns Geographische Mitteilungen (Gotha) Ergänzungs-Heft, **25**, pp. 1–75.

Rohlfs, G., 1881, *Kufra. Reise von Tripolis nach der Oase Kufra*, (Leipzig: Brockhaus), pp. 1–559.

Rönneseth, O., 1982, *Gräber im nordwestlichen Tibesti (Tschad)*. Materialien zur Allgemeinen und Vergleichenden Archäologie (München), **8**, pp. 1–65.

Staewen, C. and Striedter, K.H., 1987, *Gonoa – Felsbilder aus Nord-Tibesti (Tschad)*. Studien zur Kulturkunde (Stuttgart), **82**, pp. 1–325.

Tilho, J., 1921, L'exploration du Sahara oriental. Carte du Tibesti, du Borkou et de l'Ennedi, *La Géographie* (Paris), **36**, pp. 295–318. (avec 1 carte hors texte au 1/2.000.000)

Tillet, T. and Gabriel, B., 1990, Ehi Kournéi: Ein Acheuléen-Fundplatz im Tibestigebirge, *Berliner Geographische Studien*, **30**, pp. 277–300.

Vincent, P., 1963, Les volcans tertiaires et quaternaires du Tibesti occidental et central (Sahara du Tchad). *Mémoires du Bureau de Recherches Géologiques et Minières*, **23**, pp. 1–307.

Vincent, P., 1969, Le massif granitique de Yedri (Tibesti). *Bulletin de l'Institut Equatorial de Recherches et d'Etudes Géologiques et Minières*, **14**, pp. 89–95.

CHAPTER 3

The geomorphology and river longitudinal profiles of the Congo-Kalahari Watershed

Tyrel J. Flügel

*Department of Environmental and Geographical Science,
University of Cape Town, South Africa
Department of Earth Science, Institute for Water Studies,
University of Western Cape, South Africa*

Frank D. Eckardt

*Department of Environmental and Geographical Science,
University of Cape Town, South Africa*

Woody (Fenton) P.D. Cotterill

*Geoecodynamics Research Hub, Department of Earth Sciences,
University of Stellenbosch, South Africa*

ABSTRACT: The landscape of central and southern Africa is dominated by two basins: the low lying Congo Basin and the elevated Kalahari Basin (consisting of the Cubango and Zambezi basins). A remarkable feature of south-central Africa is that headwaters of these two large rivers, which flow in opposite directions, do not originate in a mountain range. Rather the Congo-Kalahari Watershed is a composite landscape feature that includes mountainous, wetland, plateau and flat land with the headwaters of the rivers being interleaved. This study makes use of Landsat 7 ETM+ imagery and SRTM data to investigate the geomorphology of the sub-continental region of the Congo-Kalahari Watershed. We present a series of longitudinal profiles of selected rivers, some for the first time, from the Congo-Kalahari Watershed as well as the topographic profile of the watershed itself. The topographic evidence suggest that the both the Congo and Kalahari Basins are multistage landscapes having zones of late Neogene modification and rejuvenation adjacent to landforms that appear to be highly stable and of likely Early Cenozoic in origin. This is evidenced both in the river profiles themselves as well as the topographic cross-section of the watershed.

3.1 INTRODUCTION

3.1.1 Africa's mega-geomorphology

Once the heartland of Gondwana, Africa is the only continent to straddle the Equator and has experienced multiple periods of tectonics, subsidence, uplift, denudation and deposition that modified its landsurface, with modification still ongoing (de Wit *et al.*, 2008; Torsvik and Cocks, 2009). Accounting for ca. 20% (30.3 million km²) of the

Earth's land area the continent incorporates the West African, Congo, Kalahari and Tanzania cratons (amongst others) and the East African Rift System (EARS). While there is some debate with regard to the mechanism driving the overall topography of Africa, (for example, see Nyblade and Roberts, 1994; Burke *et al.*, 2003; Doucouré and de Wit, 2003; de Wit, 2007; Burke and Gunnell, 2008) it is broadly agreed that major elements of Africa's topography, such as some rivers, erosional surfaces and depositional basins, have their origins in the Cretaceous or even earlier (for example, see King, 1953; Partridge and Maud, 1987; Doucouré and de Wit, 2003; Burke and Gunnell, 2008). Yet the actual landforms and the extent of their inheritance (if any) is a field of ongoing and actively increasing research (i.e. Roberts *et al.*, 2012; Guillocheau *et al.*, 2015; Flügel *et al.*, 2015; Agyemang *et al.*, 2016).

The cause of Africa's globally unique mega-geomorphology, characterised by its high plateaus and flat lowlands, has been described and debated for close to a century now (e.g. du Toit, 1933; Veatch, 1935; Dixey, 1938). Africa's elevation has a bimodal distribution consisting of two flat lands: one of low elevation (averaging 400 m a.s.l.) and one of high elevation (approximately 1000 m a.s.l.), whereas the rest of the world's continents displays a more unimodal elevation distribution (where that is low elevations are the majority). On a sub-continental scale, the low and high flat lands of bimodal Africa are not homogenous but rather comprise several basins and swells (Holmes, 1965). Low Africa has seven basins and high, southern Africa is comprised of three basins; interestingly high, eastern Africa has no large, notable basins, consisting predominantly of a large swell with the EARS being the primary cause of high elevation in east Africa (Holmes, 1965). The continental scale geomorphic features of the Congo Basin, Kalahari Plateau and the EARS are important factors in the evolution of Africa's south–central geomorphology. However, the exact nature, timing and magnitude of their role in determining the development of the river networks of the region requires a better understanding of these features at a temporal and spatial resolution that is unavailable at present. It is thought that since the break-up of Gondwana, Africa's continental drainage systems have undergone substantial rearrangements (Moore and Larkin, 2001; Goudie, 2005; Roberts *et al.*, 2012). However, the timing and spatial location of these rearrangements are poorly known.

Initial physiographic studies of Africa, central and southern Africa in particular, focused on how these flat surfaces were formed in low and high Africa, with cycles of erosion and uplift being proposed (du Toit, 1933; Maufe, 1935; Veatch, 1935; Dixey, 1944; Robert, 1946; King 1951; Cahen, 1954). This led to attempts to identify and correlate common erosion surfaces in an effort to determine the sequence, duration, causes and magnitude of these periods of erosion and uplift, culminating in King's well known African surfaces (King, 1951; 1962; Partridge and Maud, 1987). However, given Africa's large area, the rugged terrain and the multiple countries involved much of these earlier geomorphic observations were made on measurements and observations derived from multiple methodologies, by a number of agencies and of various accuracies. By necessity many studies focused on individual features or made generalised regional observations based on the best available data. It is only recently, with advance in earth observation techniques and data, that studies of earth surface process have been able to switch from focus from individual features to regional scale studies (see, Jarvis *et al.*, 2004; Iwahashi and Pike, 2007; Alexander, 2008). This has allowed the revisiting of large scale geomorphological problems that had previously been hindered by data issues. This study makes use of remotely sensed data to explore and detail selected rivers of the both the central Congo Basin and the southern Cubango (Okavango) and Zambezi Basins and to investigate the regional geomorphology of south-central Africa.

3.1.2 South-Central Africa: The Congo-Kalahari Watershed

Early on in the study of Africa's bimodal topography the drainage divide of the Congo and Zambezi Basins was suggested as the boundary between low, central Africa and high, southern Africa (Dixey, 1938). First mapped in detail by Steel (1917), the Congo-Zambezi watershed separates southerly flowing rivers of the Kalahari Plateau from the northerly flowing rivers of the Congo Basin (Dixey, 1943). Later studies extended the Congo-Zambezi watershed through the addition of related watersheds, the Congo-Zambeiz forming the central part of King's (1962) continent spanning West-East drainage divide. This study follows King's (1962) grouping of the Cubango (Okavango) and Zambezi basins as contemporise, given their shared geomorphic evolution (Moore and Larkin, 2001). As both the Cubango and Zambezi river systems occur predominantly on the Kalahari Plateau they are referred to collectively as the Kalahari Basin. Therefore the swell (basin divide) separating the Congo and Cubango–Zambezi Basins is termed the Congo-Kalahari Watershed (CKW). For the rest of this study the CKW will be used when referring to the drainage divide that separates the westerly flowing Congo Basin (CB) from the easterly flowing Kalahari Basin (KB) as shown in Figure 1.

The CKW separates the mostly northern flowing tributaries of the world's second largest river by discharge, the Congo (annual average discharge of 1250 109 m^3) from southern flowing tributaries of the Zambezi, the twenty first largest (annual average discharge of 220×109 m^3) (Meade, 1996; Gupta, 2007). The CKW comprises principally the Angolan Highlands (in the west), the northern limit of the Kalahari Plateau (centre) and the margins of the Western Branch of the EARS. The western CKW is bifurcated by the headwaters of the Cuanza river system that have eroded along the granite–Kalahari sand boundary resting approximately 190 km south of the general line of the CKW (Wellington, 1955). East of the Cuanza basin, the watershed separates the Zambezi headwater tributaries of the Lungwabungu, Luena and Chavuma rivers from the east–flowing headwaters of the Kasai River (Wellington, 1955). Further east, near 24°E, the rise of the watershed on the Congo side is abrupt whereas the Zambezi side is very flat, showing a gradual rise (Steel, 1917). Close to 25°E the Zambezi tributaries sources of the Lunga and Kabompo rivers are within 3 km of one another, yet the Kabompo headwaters flow due south while the Lunga headwaters flow due north, similar to the Zambezi headwaters, toward the CB (Steel, 1917). Near the upper Congo (Lualaba) river of the CB, the major landforms are a plateau complex, the result of the south–west extension of the Western Branch of the EARS (De Dapper, 1988; Gumbricht *et al.*, 2001; Chorowicz, 2005). The region east of 31°E is characterised by the rough topography of the Luangwe Rift Valley, with the Luangwe River forming part of the Zambezi drainage. On the Congo side, several short, streams flow north–west off the rift flank to form part of the Chambeshi River sub-catchment (Figure 1).

A definitive example of a major river capture of KB river by a CB river near the CKW is Kasai with its long eastward flowing headwaters flowing parallel to the CKW. This eastward flow is in contrast with the general northward of other CB rivers in the region; the location of the elbow of capture indicates the beheading the Chavuma River of the Zambezi system (Figure 1; King, 1951; Wellington, 1955). In this west-central region of the CKW (the Kasai and Zambezi river sources) extreme planation, lack of definitive flow directions and occurrence of wetlands in many headwater regions on the CB side of the divide (i.e. Lulua and upper Congo) may indicate stream diversions away from the KB (Veatch, 1935; King, 1951). Westward of the Congo–Zambezi, the headwaters of the CB rivers (in Angola and Democratic Republic of the Congo) have deeply incised valleys suggesting river rejuvenation that is presently moving the CKW

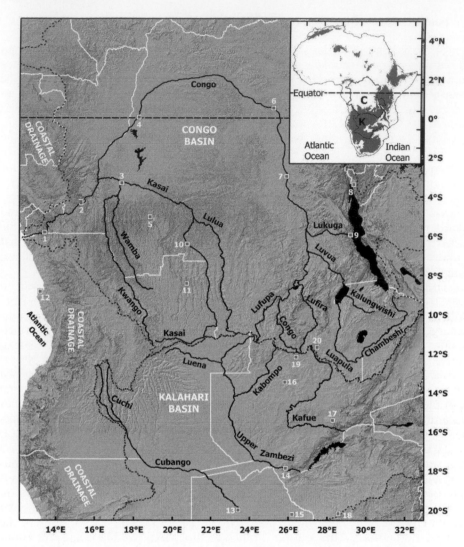

Figure 1. Map of south-central Africa indicating the studied rivers. Heavy black lines indicate the studied rivers, major water bodies are shown in black and political boundaries in white. The dotted lines indicate the drainage divides of the Congo and Kalahari Basins and coastal river systems, as determined in the study. Towns and settlements of interest have been shown: 1–Matadi; 2–Brazzaville/Kinshasa; 3–Bandundu; 4–Mbandaka; 5–Kikwit; 6–Kisangani; 7–Kindu; 8–Bujambura; 9–Kalemie; 10–Tshikapa; 11–Lucapa; 12–Luanda; 13–Maun; 14–Livingstone; 15–Nata; 16–Kasanka; 17–Lusaka; 18–Bulawayo; 19–Solwezi; 20–Lubumbashi. Inset: Congo (C) and Kalahari (K) basins (black outline.) The distribution of topography above 1000 m a.s.l. is shown by dark grey shading. Low Africa consists of the western, northern and central regions (white) while high Africa is formed by the southern and eastern regions (dark grey).

southward (King, 1951). It has been suggested that the Congo–Zambezi watershed (the central and eastern zone of the CKW), was further north of its present position forming and that is was formed by warping of the sub-Kalahari surface during the Cretaceous (Dixey, 1943; Wellington, 1949) or late Cenozoic (King, 1951). However, the causation, timing and magnitude of this change is debated (for example Roberts *et al.*, 2012; Guillocheau *et al.*, 2015; Key *et al.*, 2015).

The CKW is thus a composite landform, where the interaction between old, inherited landforms and younger, dynamic landforms (the EARS) can be investigated. An improved characterisation of the divide may aid in the understanding of the origin and development Africa's mega-geomorphology. This study aims to achieve a better understanding of the regions land surface by mapping the river longitudinal profiles of the adjacent drainage basins and characterising the CKW itself.

3.1.3 The Congo and Kalahari basins

The entire drainage network of the CB has been active since the late Cretaceous (Robert 1946; Cahen, 1954; Goudie, 2005) with Deffontaines and Chorowicz (1991) suggesting a high degree of drainage inheritance and superimposition originating in (at least) the early Cenozoic. Flügel *et al.* (2015) propose that many of the rivers of the CB underwent major changes in the Neogene, attaining their approximate present day patterns between the Mid Miocene to Pleistocene. The Western Branch of the EARS runs the entire length of the CB's eastern margin and subsequently the eastern drainage has been heavily influenced by the neotectonic activity of the EARS (Flügel *et al.*, 2015). The Basin's highlands are dominated by the northern extensions of the Kalahari Plateau (ca. 1100 m a.s.l.). At the centre of the CB are late Neogene and Quaternary alluvium mantling overlying thick continental sediments and it is thought these continental sediments, originating from erosion of the surrounding periphery, have been accumulating in the basin since Gondwana (Giresse, 2005; Runge, 2007).

The KB corresponds to the Cubango and Kalahari basins of Holmes (1965) and EARS forms the eastern boundary in the form of Lake Malawi and the Luangwa Rift. The western Cubango River system is hydrologically separate from the more extensive Zambezi River system. Many of the present day southern tributaries of the KB drainage system are fossil drainages, rarely having flowing water (Dixey, 1955). The present day drainage of the KB is thought to have originated on the surface of the Kalahari Group, and has been and currently is being superimposed on the underlying rocks as evidenced by rapids and gorges in resistant rocks along the Zambezi and Cubango rivers (Wellington, 1949; Nugent, 1990; Haddon and McCarthy, 2005). Wellington (1949) classed the ca. 2750 km Zambezi River into the Upper Zambezi (from the CKW to ca. 100 km below Victoria Falls), the Middle Zambezi and the Lower Zambezi (on the Mozambique Plain). The longitudinal profile of the Upper Zambezi river is concave (Nugent, 1990) with its course determined by the surface of the Kalahari Group (Wellington, 1949). The subsidence of the KB during the Cenozoic in combination with uplift of its margins contributed to drainage disruption leading to the separation of links between the Limpopo and Cubango river systems and the Cuando and Zambezi–Luangwa river systems (Partridge and Maud, 2000; Moore and Larkin 2001; Haddon and McCarthy, 2005). It has been suggested that doming (diameter ca. 2000 km) in southern Africa is cause for the present day drainage of the KB, having affected the Cubango and Kwango rivers and the headwaters of the Zambezi and Congo (Moore and Blenkinsop, 2002; Goudie, 2005). Similar to the CB the rivers of the KB have their origins in the Cretaceous, with Haddon and McCarthy (2005) proposing a possible model of drainage evolution of the KB from the Late Cretaceous to the early Cenozoic during the deposition of the Kalahari Group sediments. During the Cretaceous the drainage system was dominated south–east orientated rivers, draining through a main channel that is represented by the present day Limpopo River (Haddon and McCarthy, 2005). Following a series of subsidence of the central KB, drainage reversal occurred resulting in the formation of interior drainage with the subsequent deposition of the Kalahari Group sediments (Haddon and McCarthy, 2005). It was during the Cenozoic that this internal drainage was reconnected to the Indian Ocean through capturing of the Upper Zambezi by the Middle Zambezi (Nugent, 1990). Based on alluvial evidence and the Upper and Middle Zambezi both having concave profiles, Nugent (1990) suggests this capture was the result of over topping of an Upper Zambezi lake that lead to the joining of these two river profiles during the middle Pleistocene. By comparison, Moore and Larkin (2001) favours the capture of the Upper Zambezi by the westerly headward migration of the Middle Zambezi through the Miocene. During the Miocene the Chambeshi flowed south via the Kafue river into an internal drainage basin, with this internal drainage being captured by the Middle Zambezi in the late Pliocene (Moore *et al.*, 2012). Tectonic activity (associated with the EARS) lead to drainage rearrangements the early Pleistocene saw resulting in the Kafue capturing Lufira draianage (CB) and the Chambeshi being captured into the CB drainage, with the modern day drainage being established in the middle Pleistocene (Moore and Larkin, 2001). It is thus likely that the fluvial systems in the eastern CKW have undergone significant changes during the Plio–Pleistocene.

Most of the studies of the regions fluvial development have focused on drainage patterns (e.g. Cahen, 1954; Deffontaines and Chorowicz 1991; Moore and Larkin, 2001; Stankiewicz and de Wit, 2006; Runge, 2007; Key *et al.*, 2015). The focus on drainages in planform is most likely due to the quality of topographic data and remoteness of much of the study region which has made the determination of accurate longitudinal profiles difficult (e.g. Robert, 1946). The filling in of this knowledge gap with regards to longitudinal profiles in the region is the main aim of this paper.

3.1.4 Longitudinal river profiles

It may be possible to better understand Africa's Neogene development by at evolution of the regions fluvial geomorphology. River may act as recorder of broader landscape dynamics including base level change (absolute and local), denudation and localised and regional tectonics, all of which affect rivers horizontal (the course of a river itself and its associated catchments river patterns) and vertical (longitudinal profiles and valleys) morphology. The extent and length of time recorded by rivers is the result of the interplay of exogenic and endogenic factors.

Driving mechanisms that affect the development of a river's longitudinal profile are varied and include the resistance of the channel bed to erosion, the amount and type of sediment moving through the system, the tectonic setting as well as autogenic fluvial process (Leopold *et al.*, 1964; Tooth *et al.*, 2002; Whipple, 2004). Under the idealised conditions of a homogenous surface and river discharge increasing downstream, a rivers longitudinal profile tend towards a concave shape (Schumm, 1977). A concave profile represents an equilibrium, where erosion and deposition are in balance, and the river may be termed graded (Hack, 1960; Schumm, 1977). Alluvial rivers are more readily associated with concave profiles whereas many bedrock rivers deviate from the smooth concave by having both flatter and steeper zones (Leopold *et al.*, 1964; Sinha and Parker, 1996). The steeper zones, which start at knickpoints, may be the result of resistant rock outcrops, tectonic activity, sudden changes in discharge or stages of valley development such as active headwater erosion (Leopold *et al.*, 1964).

The altering of a graded river's base-level (a zone of decreased river slope and energy) results in dis-equilibrium, leading to changes in slope, channel pattern, width or roughness that persists until equilibrium is once again attained (Hack, 1960; Schumm, 1977; Leopold and Bull, 1979). However, the frequencies of disturbances is often greater than what a river can accommodate, a graded stream may be more should rather be regarded as being in dynamic–equilibrium as opposed to a true steady state (Hack, 1975; Schumm, 1977). As a river's gradient controls the morphological formations and spatial distribution of fluvial features, the noting of base-levels, both local base-level (i.e. other rivers and water bodies or resistant lithologies) and ultimate base-level (ocean), may indicate broader controls acting on the landscape (Leopold *et al.*, 1964; Whipple, 2004; Brierley *et al.* 2006). Any changes in the ultimate base-level, either through continental uplift (isostatic change) or by sea-level drop (eustatic change), have an immediate effect on drainage systems (Schumm, 2005) although the speed at which such effect ramify within a drainage system is variable. The increase of the river's length, from its source to mouth, results in an overall increase in the speed, erosive powers and carrying capacity of the river resulting in incision that migrate upstream and consequent increased rates of landscape denudation (Hack, 1960; Leopold *et al.*, 1964; Schumm, 1977; Chorley *et al.*, 1984). While ultimate base-level stops downward erosion, local base-levels only retard downward erosion (it is a matter of the time scale involved) (Tooth and McCarthy, 2002). Where a river has only a single base level (sea level) it will, over time, achieve an upward concave profile but if there are multiple persistent base levels a river may consist of a series of connected profiles tending to concavity (Whipple, 2004; Schumm, 2005). These base-levels may be due to a change in underlying rock resistance, river rejuvenation or may be due to the presence of a lake, dam, wetland or even a floodplain (Schumm, 2005).

To date there is limited publically available, spatially-referenced information characterising the whole longitudinal of rivers of the CB and KB and cross-section profiles of the CKW. This study fills this knowledge gap for selected rivers of the CKW and details the geomorphology of the CKW.

3.2 MATERIALS AND METHODS

River courses and longitudinal profiles were derived from Landsat 7 Enhanced Thematic Mapper Plus (ETM+) and Shuttle Radar Topography Mission (SRTM) imagery. The multispectral Landsat 7 ETM+ imagery provided the optical context and horizontal accuracy while the SRTM imagery provided data the elevations of the surrounding landscape and rivers. All of the geospatial data, unless otherwise stated, was referenced to the World Geodetic System 1984 (WGS84) ensuring that all of the data was spatially compatible and avoided issues of reinterpolation, which may introduce anomalies (Guth, 2009). During the analysis stages some of the data was projected to allow for measurements using the more relevant zone of the Universal Transverse Mercator (UTM) coordinate system.

3.2.1 Data used

Landsat 7 ETM+ imagery was used due to its extensive coverage and public availability, its medium ground spatial resolution (30 × 30 m pixel) and high degree of processing, orthorectified and proven use (GLCF Landsat Technical Guide, 2004;

Tucker *et al.*, 2004). To aid river digitisation three bands, green (band 2), near infra-red (band 4) and mid infra-red (band 7), were used to make false-colour composites that provided a high visual contrast between water and vegetation (bands 2 and 4) and bare soil/rock (band 4 and 7) (GLCF Landsat Technical Guide, 2004; Lillesand *et al.*, 2004). Given that Landsat 7's average horizontal displacement accuracy is often better than the stated 50 m root-mean-square error (Tucker *et al.*, 2004) Landsat imagery is a good source to ensure locational accuracy when digitising rivers at a regional scale. Landsat imagery was downloaded from the Global Land Cover Facility (GLCF Earth Science Data Interface, n.d.) with image selection criteria being minimum cloud cover and a capture data of circa 2000 to ensure a close temporal match to the SRTM data, which was captured in February 2000. Given Landsat's extensive processing, the three bands only needed to be combined to create the false-colour composite scenes. ERDAS IMAGINE 9.1 was used to produce the false-colour composites, where band 7 was displayed as red, band 4 as green and band 2 as blue. In the resulting false-colour composite images, water and saturated soils appeared in shades of blue, vegetation in shades of green and bare soil/rock and other hard surfaces (such as roads) in shades of pink. Each scene was colour-contrasted to enhance the visual differences between water, vegetation and hard surfaces so as to allow for easier differentiation of the river channel and its features.

For the longitudinal profiles and regional geomorphology SRTM data was used. Rather than the original NASA version of the SRTM, this study used the digital surface model (DSM) produced by the Consultative Group on International Agricultural Research (CGIAR). The CGIAR's Consortium for Spatial Information (CSI) processed the NASA released SRTM 3 arc second (90 × 90 m pixel) in order to resolve issues, including data holes and pixel shifts, producing several versions of thee SRTM DSM (see Gamache (2004), Jarvis *et al.* (2004) and Reuter *et al.* (2007) for further details). This study made use of CGIAR-CSI's third version of the SRTM (SRTMv3) (Jarvis *et al.*, 2006) to determine the river longitudinal profiles and the profile of the CKW. The effective on-the-ground pixel resolution of the SRTMv3 is 90 m and as data voids were limited in central and southern Africa and little void filling was required for the study region (Jarvis *et al.*, 2004; Reuter *et al.*, 2007). A coarser 250 m pixel resolution DSM (SRTM250 m) was used to produce the elevation maps and large scale cross sections; this SRTM250 m being the fourth iteration of the DSM (Jarvis et al., 2011). As the elevations of the SRTM are radar based, possible radar-landscape interactions (relief displacement, radar shadow and noise due to the inclusion of non-topographic) may affect the quality of the DSM in a non-systematic manner (Lillesand *et al.*, 2004; Nelson *et al.*, 2009). These effects can be seen as elevation peaks or sinks, or increased elevations of flat regions and decreased elevations of hill and mountain tops (Jarvis *et al.*, 2004; Ludwig and Schneider, 2006). Despite these effects the overall absolute vertical accuracies for Africa are an absolute vertical error of 5.6 m and a relative vertical error of 9.8 m (at 90% error limits) (US Geological Survey (n.d.); Rodríguez *et al.*, 2006) with absolute horizontal location accuracy for Africa (at 90% error limits) being 11.9 m error (Rodríguez Rodríguez *et al.*, 2006). The CGIAR-CSI SRTM did not require further processing for this study, with ERDAS 9.1 used to convert the files into a single IMAGINE (.img) rasters and hill shade creation for both SRTM resolutions. This was done by reprojecting the DSMs from WGS84 to a Robinson UTM projection (UTM Zone 33 south, being the middle zone of the sub-continental study region, was chosen) and nearest neighbour resampling used. The vertical scale was set as three and the solar azimuth and elevation were set to 45°. The hill shaded relief output was then reprojected back to same coordinate system as the original SRTMv3, and while double reprojection may introduce minor horizontal offsets, as the hill shaded relief was interpretively these offsets were considered negligible.

3.2.2 River digitisation and profile creation

Rivers were digitised from source to end, utilising a method similar to McFarlane and Eckardt (2007). The false-colour composites, SRTMv3 and hillshaded images were used simultaneously by linking the ERDAS viewer windows geographically, this allowed the same location to be indicated on all three images regardless of individual viewers scales. Digitisation (using the profile tool) followed a two-step approach: firstly, the river course was digitised following the lowest valley elevation as indicated on the SRTMv3, followed by a second, refinement step where horizontal course adjustment were undertaken using the river channel as seen in the Landsat image. This adjustment was done so as to be as close to the thalweg as possible whilst still being in the stream as seen in Landsat. River courses as seen in the Landsat imagery were given primacy over the SRTM equivalent due to Landsat's more extensive quality control methods, its better horizontal spatial accuracy and 30 m pixel resolution compared to SRTM's 90 m resolution and topographic issues. This second step ensured the horizontal fidelity (latitude and longitude) of the digitised river location, allowing the same river course shapefiles to be used should a better DSM become available.

The ERDAS Profile Tool extracts all of the pixel elevations falling on the straight line between the two manually placed nodes. Complex river course necessitated a higher density of points (up to a point every ~30 m) while simple courses (straight stretches) had lower point densities (a point every ~1 km). In regions of high course sinuosity and limit elevation change (i.e. floodplain zones), meanders with a wavelength of less than 50 m were straightened out, while larger meanders (wavelength of ca. 100 m or more) had their average curve followed; this was done to decrease digitisation time. Given the highly dynamic depositional environments of most floodplains, with meanders being of a short-lived nature, it is unlikely that accurate digitisation of these meanders would significantly improve the accuracy of the data or detract from the geomorphologic interpretations of the river systems. In areas of incised meanders (indicated by a large change in elevation of river surface and surrounds) this method was not followed as it would lead to artificial knickpoints and so meanders were accurately traced. As all the imagery was in the WGS84 geographic coordinate system, river distanced were measured in decimal degrees and pixel number. The WGS84 coordinate system was used as most of the studied rivers crossed several degrees of latitude and longitudes and thus using a single projection may have introduced spatial errors. A total of 16 river long profiles and courses were manually digitised, with these rivers forming part of the CKW, as seen in Figure 1. Knickpoints where recorded using a combination of the elevation data and the Landsat imagery. The digitisation of the CKW and coastal watersheds the same but an aspect raster was substituted for Landsat imagery as it allowed for better discrimination of hill slope direction.

The output of digitisation was a table of elevation data per a pixel, each pixel record having an associated longitude, latitude amongst other attributes. This table was converted into a spreadsheet allowing for a river course and profile geodatabase. The elevation point data was used to create point and polyline shapefiles in ArcGIS 10. The result was two shapefiles per a river: point shapefiles (which had elevation attribute data) and polyline shapefiles (which allowed for easier mapping and spatial analysis). To determine river lengths in kilometres, river polyline shapefiles were duplicated, and projected to a south Universal Transverse Mercator (UTM) zone that was closest to the rivers mid-point, allowing for metric length calculations. As not all meanders were digitised, river lengths should be considered as minimum distances. This approach may result in river lengths that differ from those reported elsewhere, but the fractal properties of river networks (e.g. Tarboton *et al.*, 1988) and effects of raster resolution on length measurements should be kept in mind.

3.3 RESULTS

A total ca. 19,700 km of river course was digitised and 380 knickpoints identified (Table 1 and Figure 2). In the CB ten river profiles consisting of ca. 13,300 river kilometres were digitised and 259 knickpoints identified (four of which were large dams). The CB's three longest rivers are the Congo (ca. 4100 km), Kasai (ca. 2100 km) and the Chambeshi-Luapula-Luvua system (ca. 1700 km) (Table 1). Six rivers were digitised in the KB, covering ca. 6300 river kilometres, and 121 knickpoints identified (two of which were large dams). For the KB, the Cubango (ca. 1700 km), Kafue (ca. 1500 km) and Upper Zambezi (ca. 1500 km) rivers are the top three longest rivers (Table 1). Figure 2 shows the individual river profiles at the basin scale, with knickpoints being indicated, while Figure 3 displays the river at the same scale for the CKW to allow a comparison between the CB and KB. As can be seen in Figure 2 several of the long profiles are noisy, this noise (seen as spikes and holes) being the result of a combination of

Table 1. Key characteristics of digitised longitudinal profiles. Pixel no. indicates the total number of 90 m SRTM pixels that form a river course. River length (Length km) was calculated by projecting the relevant river shapefile to UTM, using the zone closest to each river's middle point and calculating length.

River name	No. of knickpoints	Pixel No.	Length (km)	Start elevation (m a.s.l.)	Finish elevation (m a.s.l.)	Total drop (m)
Congo basin
Chambeshi-Luapula-Luvua	30	16,584	1779	1576	556	1020
Congo	64	42,715	4150	1526	0	1526
Kalungwishi	9	2686	291	1592	921	671
Kasai	47	20,718	2064	1348	272	1076
Kwango	18	14,661	1465	1395	292	1103
Lufira	11	6558	681	1415	562	853
Lufupa	21	5168	530	1485	562	854
Lukuga	8	3283	338	767	547	220
Lulua	30	11,290	1137	1225	381	844
Wamba	21	8869	922	1183	312	871
Kalahari basin
Cubango	55	17,074	1708	1834	937	897
Cuchi	14	5118	524	1757	1191	566
Kabompo	4	6568	653	1495	1031	464
Kafue	14	14,744	1529	1375	369	1006
Luena	10	4313	430	1403	1063	340
Upper Zambezi	24	12,600	1522	1464	505	959

radar-landscape interactions, and for the CB especially dense tree cover near the narrower sections of river channel.

In the CB (Figure 2 top) the Congo River acts as the base-level for all of the studies rivers in the CB. The profile of the Congo River itself can be divided into three sections: (1) the steep upper profile from source to the juncture with the Lukuga

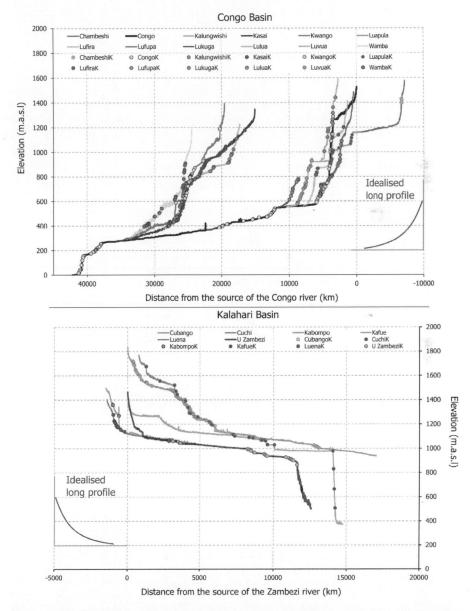

Figure 2. The longitudinal profiles of studied rivers (coloured polylines) and location of knickpoints (coloured dots) of the CKW. The rivers are shown in their relative location to one another. (Top) The long profiles of the Congo and (Bottom) Kalahari with knickpoints indicated. The profiles shown indicate the actual SRTMv3 elevation data (i.e. data has not been averaged). Note the different horizontal scales used and the schematic idealised river long profile inset.

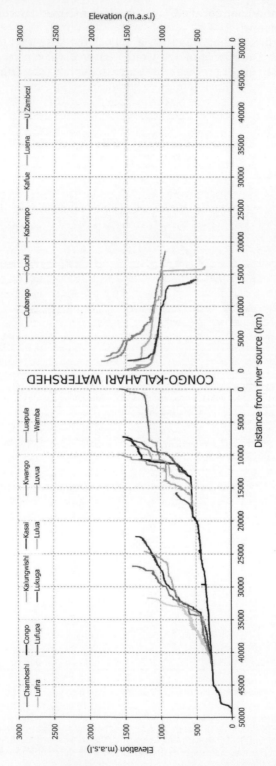

Figure 3. A comparison of the studied rivers on the same scale (horizontal and vertical) to allow for comparison of the longitudinal profiles from the two basins. The CB rivers (left) have a more complex longitudinal characteristics compared to the rivers of the KB (right). Rivers in both systems (except the Lukuga river of the CB) have their origin in the CKW with source elevation ca. 1400–1500 m a.s.l.

River; (2) a flat, gentle gradient profile (3) and the lower Congo River downstream of Kinshasa which is steep with a high density of knickpoints. The upper tributaries (Kalungwishi, Lufira, Lufupa and Lukuga), apart from the Chambeshi-Luapula-Luvua, have relative steep concave gradients with much of the change in elevation over short distances (Table 1 and Figure 2). The Chambeshi-Luapula-Luvua profile is a composite river with the Chambeshi appearing graded, the Luapula having two distinct steps and the Luvua having a steep gradient. It should be noted that the Chambeshi-end Luapula-source is the Bangweulu Swamp and the Luapula-end Luvua-source is Lake Mweru and thus the Chambeshi-Luapula-Luvua has two well established base-levels before it flows into the Congo River.

The lower Congo tributaries (Kasai-Lulua and Kwango-Wamba) profiles differ from the upper tributaries. The lower tributary profiles trend to having steep convex upper regions with a convex middle section followed by graded lower reaches. The division of these three sections corresponds broadly to the following manner: upper concave reaches are ca. 900 m a.s.l. for the the Kasai, Lula and Kwango, and ca. 600 m a.s.l for the Wamba. The middle convex regions of the Kasai, Lulua and Kwango are between ca. 900 and 400 m a.s.l., where most of the knickpoints are clustered; here change in profile elevation occurs over short distance. A similar pattern is seen for the Wamba between ca. 600 m a.s.l. and 400 m a.s.l. For all the upper tributaries, gentle graded profiles are seen from ca. 400 m a.s.l. until their end.

Interestingly only the Lufupa River has a knickpoint within 20 km of it juncture with the Congo River; for the rest of the studied rivers their knickpoints occur more than 20 km upstream from their juncture with the Congo River. For the CB the relationship of the knickpoint distribution (an indicator of extreme steep river gradients) does not appear to show any uniform distribution, although individual rivers do show groupings of knickpoints (Figure 3). Nor does river length appear to be the sole determinant for knickpoint number (Table 1) indicating a non-systematic control of river gradients in the CB.

In the KB the studied river profiles are, in broad terms similar (Figure 2 bottom). The Upper Zambezi's upper and middle sections form a concave, graded profile that approximates the idealised profile. The upper Zambezi's lower region (starting with a series of knickpoints) has a stepped appearance with three smaller step followed by a dramatic decrease in elevation at the Victoria Falls. The Zambezi tributaries of the Luena and Kabompo have steep upper regions but flatten out ca. 1000 m a.s.l. before entering into the low gradient zone of the Upper Zambezi. The Kafue River has three distinct regions, a graded upper and graded middle section. The lower section, where most of the Kafue knickpoints begin represents the start of the Kafue Gorge, which in modern times has been dammed as seen by the two horizontal steps starting at the 1000 pixel count mark. The Kafue continues to flow in a steep gorge after the second dam wall. The Cubango-Cuhi river systems differ from the other KB systems in that they both are steep and more convex between ca. 1500 and ca. 1200 m a.s.l. The upper Cuchi River is relatively steep with a stepped concave profile, while the upper Cubango zone is convex. The lower Cubango (downstream of the Cuchi-Cubango juncture) has a low, flat gradient.

In terms of knickpoints distribution in the KB the majority of the knickpoints are confined to upper headwater regions or occur as the river approaches elevations of 1000 m a.s.l. The Luena, Cuchi and Cubango river all show knickpoints within 20 km upstream of their junctions, at 6 km, 6 km and 11 km respectively. Additionally knickpoint number is proportional to river length (Table 1) but they do show geographic clustering at the individual river level.

When comparing the profiles of the two basins to one another (Figure 3) there are some notable trends. Firstly, there are two groups based on source elevation: those

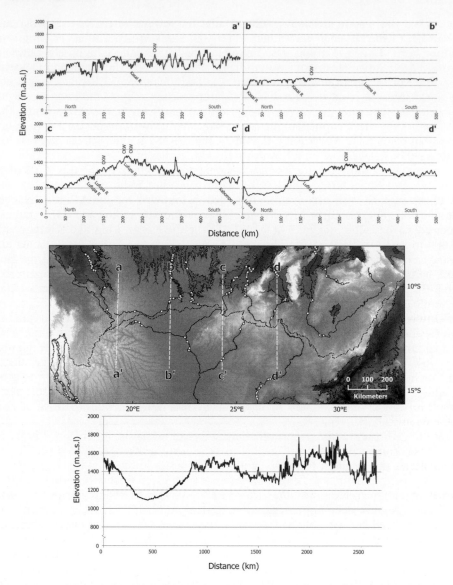

Figure 4. Topographic profiles across (top) and of (bottom) the Congo-Kalahari Watershed (middle). Top: The CKW is displays a divers topography being rugged in the west (a-a'), becoming nearly flat further east (b-b'), to have a classic watershed A shape (c-c') in the central region and have a stepped shape in the east (d-d'). Middle: The location of the profiles a to d (dashed white line), and the watersheds (dotted blackline). Heavy black lines indicate studied rivers, white circles location of identified knickpoints and broken black line the 1000 m a.s.l. contour. Bottom: The highline of the watershed itself is become more progressively rugged toward the east. Profiles across the watershed are from SRTMv250, CKW profile is SRTMv3. Note the truncate elevation (y-axis) scale of all the profile data.

rivers starting between ca. 1570 to ca. 1340 m a.s.l and those falling outside of this range (Table 1 and Figure 3). The rivers falling outside of the 1570–1340 m a.s.l. are the Cubango and Cuchi (1834 and 1751 m a.s.l. respectively) and the Lulua (1183 m a.s.l.) and the Wamba (1225 m a.s.l.). Otherwise the rivers of the CB and KB have their sources in the between 1576–1348 and 1495–1375 m a.s.l. respectively.

Those profiles with similar head water regions (in terms of elevation and topography) resemble each other more closely, even across the CKW. This is seen in the Upper Zambez, Kafue and Chambeshi-Luapula rivers, all who have headwaters in region of low ruggedness. Similarly the upper tributaries of the Congo (excluding the Chambeshi) and the Kwango and Lulua exhibit similar steep upper, headwater regions. The Kafue and Chambeshi-Lupuala have strikingly similar profiles, both with graded upper reaches and low gradient region separated by a step in the profile.

The cross-section of the CKW and the profile of the CKW itself provide the context for some of these observed features (Figure 4). The four watershed cross-sections illustrate the diversity of the watershed in the north-south direction as one moves from west to east. Cross-section a-a' (Figure 4 top) shows that the KB side of the watershed has elevations that are higher than the CB side, with the CB side have an incised northerly dipped plateau land. Cross-section a-b' shows the near lack of a watershed in the central region of the CKW. The CB side shows a series of incised river valley but their appears to be no evidence of incision in the south (KB). Cross-section c-c' illustrates an expected A-shaped watershed. Both basins see decreasing elevations away from the watershed apex, with elevations decreasing quicker on the CB side. Cross-section d-d' does not have a distinct watershed, the KB appears to be consistently 200 m higher than the CB in this region with both sides being relatively flat and few deeply incised river valley are apparent.

The actual profile of the CKW itself shows a great diversity of elevation ranges, from a low of ca. 1100 to a high of ca. 1780 m a.s.l. (Figure 4 bottom). Moving from west to east the CKW goes from a highland with incised valleys to a concave region of little topographic diversity. East of this smooth region the watershed increase in elevation, with elevations falling in the 1400–1600 m a.s.l. band over the next 1000 km. Elevations decrease to ca. 1300 m a.s.l. over the next 500 km east. At a distance of 1750 km from the western start of the CKW, this eastern section of the begins to displays large variations in elevation, with a maximum elevation of ca. 1780 m.a.s.l. and a low ca. 1275 m a.s.l. the elevations show a sequence of spikes and drops over a the eastern ca. 1000 km.

3.4 DISCUSSION

3.4.1 Longitudinal profiles of the CKW

None of the profiles display a smooth concave shape (Figure 2) suggesting that the fluvial systems of both basins are in dis-equilibrium. Further, given the differences in profiles both within the basin and between the basins (Figures 2 and 3) the cause of this dis-equilibrium is multi-faceted. For the CB the lack of knickpoints being clustered at river junctions implies that the Congo River has been the base level for the rest of the rivers for a prolonged period of time. Indeed, the clustering of knickpoints in the lower Congo suggests that the ocean as the ultimate base level is still unseen by the rest of the rivers of the CB. Similarly the distribution of knickpoints in the KB does not suggest knickpoint occurrence to be a function of changes ultimate base level in either basin. Thus knickpoint occurrence is likely to be dominantly controlled by structural or lithological factors throughout the CB and KB. Given the sub-continental extent of the CB and KB, the development of the longitudinal profiles is probably due to the interplay of several influences interacting in both enhancing and subduing development of river grade. Sinha and Parker (1996) noted that the major changes imposed on a river system may not necessarily result in significant changes in the river's longitudinal profile, as these changes may replace one driving mechanism for another. What is noticeable is that the fluvial systems of the CB and KB have undergone significant change has during the Neogene but these changes have not been uniform or constant (Figures 2, 3 and 4). However where rivers flow over large portions of the sediments of similar erodibility, such as the central CB and the Kalahari Sands of the KB (Dixey, 1943; Cahen, 1954; Haddon and McCarthy, 2005; Runge, 2007), river gradients are flatter (Figure 5).

This is seen in the KB by the profiles of the middle and lower Cubango, lower Luena and Kabompo and the Zambezi; and in the CB by the middle Congo and lower Kasai. In this region the Kalahari succession is more than 30 m deep, and in many instances

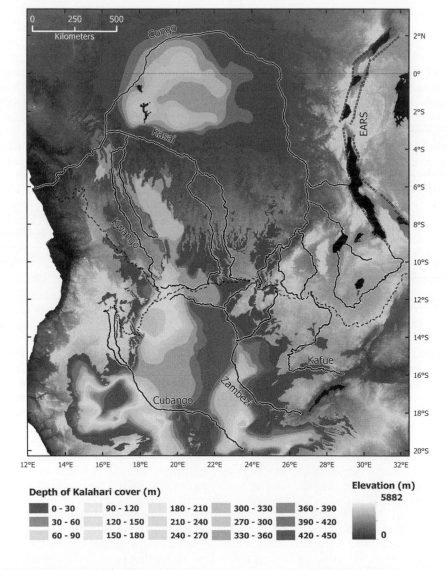

Figure 5. The extent of the Cenozoic Kalahari group across south-central Africa. Note the coverage of depths greater than 30 m in the noerthwestern KB and central CB. The East Africa Rift System (EARS) is indicated by double dashed lines, and the CKW by the single dashed lines. Studied rivers are shown in solid heavy lines with white border and major water bodies shaded black. Selected river names provided for reference. Isopach values for the Kalahari plateau from Haddon (1999) and those for the CB from Linol *et al.* (2015). Elevation data is SRTM 250 m DEM (Jarvis *et al.*, 2011).

is dominated by unconsolidated/poorly consolidated sediments (Figure 5; Linol *et al.*, 2015) and it is here that the flattest portions of both basins are found. Given the sedimentary depth it is likely that the rivers (or long reaches of the river) are able to adjust to changes in equilibrium to maintain a low gradient. Thus for these rivers any change to equilibrium in the lower river is likely to be accommodated quickly and transferred to the headwater regions. This may explain the relatively steep headwaters of these rivers.

3.4.2 Western (16°–25°E) CKW

The western CKW, from the outer watershed of the Cubango-coastal watershed to 25°E, appear to be a stable watershed. There is little evidence for landscape rejuvenation in the region (Figure 4), with the part of the western section displaying aan A-shaped watershed (Figure 4, c-c'). However between 20–23°E (Figure 4, b-b') there is not apparent watershed in this region and it likely to be very dynamic in terms of fluvial driven processes. West of 20°E the watershed become entrenched in the Angolan Highlands and this is likely to be the oldest and least modified part of the watershed, being older than those sections east of it. The main changes to the watershed in this region are likely to those associated with aggressive headward erosion by the coastal drainage of the Cuanaza coast river system.

3.4.3 Eastern (25°–33°E) CKW

The eastern CKW, east of 25°E, displays modification of a pre-existing watershed by the extension of the EARS, this being most evidently seen in the profile of the Chambeshi-Luapula-Luvua (Figure 2) and the CKW (Figure 4, bottom, east of the 1500 km mark). The Luapula segment flows into Lake Mweru, a rift lake that was formed in the seismically active as a result of the south-western extension of the EARS's Western Branch (Gumbricht *et al.*, 2001). The stepped nature of the Chambeshi-Luapula-Luvua profile and its multiple direction changes indicates the river flowing through a two rift block. This in combination with the steep profiles of the Kalungwishi and Lufira, which match that of the Lukuga river, further highlight the alteration of the regions drainage by the EARS. Based on ages for landforms associated with the EARS in this region, the eastern CKW has been undergoing modification since at least the late Neogene (Cahen, 1954; Wichura *et al.*, 2011; Delvaux *et al.*, 2012) with major drainage changes occurring well into the Pleistocene (Moore and Larkin, 2001; Flügel *et al.*, 2015) with changes likely to be ongoing given the regions seismicity (Mavonga and Durrheim, 2009). Despite being south-west of the Luapula, the profile of the upper Kafue near the CKW appears graded, suggestion that modification by the Western Branch has yet to impact upon the upper Kafue. Nevertheless, the eastern region can thus be considered the most modified and dynamic segment of the CKW given the elevation difference between the highest divides and valleys (Figure 4, bottom).

3.5 CONCLUSION

We have provided the first description and first analysis of the combined longitudinal profiles of selected rivers from the CK and KB. Initial analyses of the rivers show evidence for quasi-equilibrium in multiple sections of rivers in both basins, although the KB rivers display a higher degree grade. The lower base levels of the CB, along with the more varied profiles suggest that the CB is the more dynamic of the two basins and is likely to be the younger of the two systems. This suggests a continued southward migration of the CKW, especially as several of the CB rivers headwaters are at lower elevations compared to the KB. It is likely that multiple headwater captures and reversal of headwater captures has occurred in the zone of 20–23°E where the watershed is of extremely low gradient. While drainage capture in the eastern CKW is likely driven by the geodynamics related to the EARS, this low gradient section is likely to dominated by fluvial autogenic processes. Further investigation into the role of geology and tectonics in the development of the fluvial systems of the CKW is required to gain a fuller understanding of the regions geomorphic evolution. Given the pattern of knickpoint clustering on an individual river scale, but not at the basin scale, it is probable there is a non-fluvial control on these identified knickpoints. This clustering of knickpoints does not appear to be common within the Kalahari sedimentary cover, suggesting further analyse is required to determine the role relative role of bedrock channels and alluvial channels as determinants (Sinha and Parker, 1996; Whipple, 2004) in evolution of the rivers of the CKW. As knickpoints are critical nodes in river evolution, characterisation of knickpoints would be an effective manner to further investigate the fluvial development, and by extension landscape, evolution of the region.

ACKNOWLEDGEMENTS

Tyrel Flügel acknowledges the following for their financial support that enabled this study: The National Research Foundation (South Africa), the John Ellerman Foundation, the ERANDA Foundation, the Harry–Crossley Foundation, as well as the additional support from the University of Cape Town Doctoral Scholarship fund.

REFERENCES

Agyemang, P.C.O., Roberts, E.M. and Jelsma, H.A., 2016, Late Jurassic-Cretaceous fluvial evolution of central Africa: Insights from the Kasai-Congo Basin, Democratic Republic Congo. *Cretaceous Research*, **67**, pp. 25–43.

Alexander, D.E., 2008, A brief survey of GIS in mass-movement studies, with reflections on theory and methods. *Geomorphology*, **94**(3–4), pp. 261–267.

Brierley, G., Fryirs, K. and Jain, V., 2006, Landscape connectivity: the geographic basis of geomorphic applications. *Area*, **38**(2), pp. 165–174.

Burke, K. and Gunnell, Y., 2008, The African erosion surface: a continental-scale synthesis of geomorphology, tectonics and environmental change over the past 180 million years. *Geological Society of America Memoir*, **201**, pp. 66.

Burke, K., MacGregor, D. and Cameron, N., 2003, Africa's petroleum systems: four tectonic Aces in the past 600 million years. In *Petroleum Geology of Africa: New Themes and Developing Technologies*, edited by Arthur, T., MacGregor, D. and Cameron, N., Geological Society of London Special Publication, **207**, pp. 21–60.

Cahen, L., 1954, *Géologie du Congo Belge*, (Liège: H. Vaillant Carmanne), pp. 490.

Chorley, R.J., Schumm, S.A. and David E.S., 1984, *Geomorphology*, (New York: Methuen), p. 498.

Chorowicz, J., 2005, The East African Rift System. *Journal of African Earth Sciences*, **43**, pp. 379–410.

Deffontaines, D. and Chorowicz, J., 1991, Principles of drainage basin analysis from multisource data: Application to the structural analysis of the Zaire Basin. *Tectonophysics*, **194**, pp. 237–263.

Delvaux, D., Kervyn, F., Macheyeki, A.S. and Temu, E.B., 2012, Geodynamic significance of the TRM segment in East African Rift (W-Tanzania): Active tectonics and paleostress in the Ufipa plateau and Rukwa basin. *Journal of Structural Geology*, **37**, pp. 161–180.

De Wit, M.J., 2007, The Kalahari Eperiogeny and climate change: differentiating cause and effect from core to space. *South Africa Journal of Geology*, **110**, pp. 367–392.

De Wit, M.J., Stankiewicz, J. and Reeves C., 2008, Restoring Pan-African-Brasiliano connections: more Gondwana control, less Trans-Atlantic corruption. In *West Gondwana: Pre-Cenozoic Correlations Across the South Atlantic Region*, edited by Pankhurst, R.J., Trouw, R.A.J., Brotp Neves, B.B. and de Wit, M.J., Geological Society, London, Special Publications, **294**, pp. 399–412.

Dixey, F., 1938, Some observations on the physiographical development of central and southern Africa. *Transactions of the Geological Society of South Africa*, **41**, pp. 113–170.

Dixey, F., 1943, The Morphology of the Congo-Zambesi Watershed. South African *Geographical Journal*, **25**(1), pp. 20–41.

Dixey, F., 1955, Erosion surfaces in Africa; some considerations of age and origin. *Transactions Geological Society South Africa*, **58**, pp. 265–280.

Du Toit, A.L., 1933, Crustal movement as a factor in the geographical evolution of South Africa. *The South African Geographical Journal*, **16**, pp. 4–20.

Doucouré, C.M. and de Wit, M.J., 2003, Old inherited origin for the present near–bimodal topography of Africa. *Journal of African Earth Sciences*, **36**, pp. 371–388.

Flügel, T.J., Eckardt, F.D. and Cotterill, F.P., 2015, The present day drainage patterns of the Congo River system and their Neogene evolution. In *Geology and Resource Potential of the Congo Basin*, edited by de Wit, M.J., Guillocheau, F. and de Wit, M.C.J., (Berlin, Heidelberg: Springer), pp. 315–337.

Gamache, M., 2004, Free and Low Cost Datasets for International Mountain Cartography, Alpine Mapping Guild. Available at: <http://www.terrainmap.com/downloads/Gamache_final_wed.pdf> [Accessed 3 July 2013].

Global Land Cover Facility Earth Science Data Interface, n.d., University of Maryland. [online]. Available at: http://glcfapp.glcf.umd.edu:8080/esdi/index.jsp. [Accessed May 2008 – March 2009].

Global Land Cover Facility Landsat Technical Guide, 2004, University of Maryland. [online]. Available at: http://glcf.umicas.umd.edu/data/guide/technical/techguide_landsat.pdf. [Accessed 15 April 2008].

Goudie, A.S., 2005, The drainage of Africa since the Cretaceous. *Geomorphology*, **67**, pp. 437–456.

Guillocheau, F., Chelalou, R., Linol, B., Dauteuil, O., Robin, C., Mvondo, F., Callec, Y. and Colin, J.P., 2015, Cenozoic Landscape Evolution in and Around the Congo Basin: Constraints from Sediments and Planation Surfaces. In *Geology and Resource Potential of the Congo Basin*, edited by de Wit, M.J., Guillocheau, F. and de Wit, M.C.J., (Berlin, Heidelberg: Springer), pp. 271–309.

Gupta, A., 2007, Introduction. In *Large Rivers: Geomorphology and Management*, edited by Gupta, A., (New York: John Wiley and Sons), pp. 1–4.

Gumbricht, T., McCarthy, T.S. and Merry, C.L., 2001, The topography of the Okavango Delta, Botswana, and its tectonic and sedimentological implications. *South African Journal of Geology*, **104**(3), pp. 243–264.

Guth, P., 2009, Geomorphometry in MICRODEM. In *Developments in Soil Science Volume 33 – Geomorphometry: Concepts, Software, Applications*, edited by Hengl, T. and Reuter H.I., (Amsterdam: Elesevier), pp. 351–366.

Hack, J.T., 1960, Interpretation of erosional topography in humid temperate regions. *American Journal of Science*, **258-A**, pp. 80–97.

Haddon, I., 1999, *Isopach map of the Kalahari Group, at a scale of 1: 2.500.000. Pretoria*, (South Africa: Council for Geoscience).

Haddon, I.G. and McCarthy, T.S., 2005, The Mesozoic-Cenozoic interior sag basins of Central Africa: The Late-Cretaceous—Cenozoic Kalahari and Okavango basins. *Journal of African Earth Sciences*, **43**, pp. 316–333.

Holmes, A., 1965, *Principles of Physical Geology*, Revised Edition, (Thomas Nelson: London), pp. 1288.

Iwahashi, J. and Pike, R.J., 2007, Automated classifications of topography from DEMs by an unsupervised nested-means algorithm and a three-part geometric signature. *Geomorphology*, **86**, pp. 409–440.

Jarvis, A., Reuter, H.I., Nelson, A. and Guevara, E., 2006, *Hole-filled seamless SRTM data Version 3*, available from the CGIAR-CSI 90 m Database. [online] Available at: <http://srtm.csi.cgiar.org>. [Last accessed April 2006].

Jarvis, A., Reuter, H.I., Nelson, A. and Guevara, E., 2011, *Hole-filled SRTM for the globe Version 4*, available from the CGIAR-CSI SRTM 250 m Database. [online] Available at: <http://srtm.csi.cgiar.org>. [Last accessed January 2017].

Jarvis, A., Rubiano, J., Nelson, A., Farrow, A. and Mulligan, M., 2004, *Practical use of SRTM data in the tropics: Comparisons with digital elevation models generated from cartographic data*. Working Document no 198. Cali, Centro Internacional de Agricultura Tropical (CIAT): 32.

Key, R.M., Cotterill, F.P.D. and Moore, A.E., 2015, The Zambezi River: an archive of tectonic events linked to the amalgamation and disruption of Gondwana and subsequent evolution of the African plate. *South African Journal of Geology*, **118**(4), pp. 425–438.

King, L.C., 1951, *South African Scenery: A textbook of geomorphology*, Second edition—revised, (Edinburgh: Oliver and Boyd), pp. 379.

King, L.C., 1953, Canons of landscape evolution. *Geological Society of America Bulletin*, **64**, pp. 721–751.

King, L.C., 1962, *The Morphology of the Earth: A study and synthesis of world scenery*, (Edinburgh: Oliver and Boyd), pp. 699.

Leopold, L.B. and Bull, W.B., 1979, Base level, aggradation, and grade. *Proceedings of the American Philosophical Society*, pp. 168–202.

Leopold, L.B., Wolman, M.G. and Miller, J.P., 1964, *Fluvial process in geomorphology*, (San Francisco: Freeman), pp. 552.

Lillesand, T.M., Kiefer, R.W. and Chipman, J.W., 2004, *Remote Sensing and Image Interpretation*, 5th Edition, (New York: Wiley), pp. 763.

Linol, B., de Wit, M.J., Guillocheau, F., de Wit, M.C.J., Anka, Z. and Colin J.P., 2015, Formation and collapse of the Kalahari duricrust ['African Surface'] across the Congo Basin, with implications for changes in rates of Cenozoic off-shore sedimentation. In *Geology and Resource Potential of the Congo Basin*, edited by de Wit, M.J., Guillocheau, F. and de Wit, M.C.J., (Berlin, Heidelberg: Springer), pp. 193–227.

Ludwig, R. and Schneider, P., 2006, Validation of digital elevation models from SRTM X-SAR for applications in hydrologic modelling. *ISPRS Journal of Photogrammetry and Remote Sensing*, **60**, pp. 339–358.

Maufe, H.B., 1935, Some Factors in the Geographical Evolution of Sothern Rhodesia and Neighbouring Countries. *South African Geographical Journal*, **18**(1), pp. 3–21.

Mavonga, T. and Durrheim, R.J., 2009, Probabilistic seismic hazard assessment for the Deomocratic Republic of Congo and surrounding areas. *South African Journal of Geology*, **112**, pp. 329–342.

McFarlane, M.J. and Eckardt, F.D., 2007, Palaeodune morphology associated with the Gumare fault of the Okavango graben in the Botswana/Namibia borderland:

a new model of tectonic influence. *South African Journal of Geology*, **110**, pp. 535–542.

Meade, R.H., 1996, River-sediment Inputs to Major Deltas. In *Sea-level Rise and Coastal Subsidence: Causes, Consequences and Strategies*, edited by Milliman, J.D. and Haq, B.U., (Dordrecht: Kluwer), pp. 63–85.

Moore, A.E. and Larkin, P.A., 2001, Drainage evolution in south-central Africa since the breakup of Gondwana. *South African Journal of Geology*, **104**, pp. 47–68.

Moore, A.E., Cotterill, F.P.D. and Eckardt, F.D., 2012, The evolution and ages of Makgadikgadi palaeo-lakes: consilient evidence from Kalahari drainage evolution, South-Central Africa. *South African Journal of Geology*, **115**, pp. 385–413.

Nelson, A., Reuter, H.I. and Gessler, P., 2009, DEM production methods and sources. In *Developments in Soil Science, Volume 33 – Geomorphometry: Concepts, Software, Applications*, edited by Hengl, T. and Reuter H.I., (Amsterdam: Elsevier), pp. 65–85.

Nugent, C., 1990, The Zambezi River: tectonism, climate change and drainage evolution. *Palaeogeography, Palaeoclimatology, Palaeoecology*, **78**, pp. 55–69.

Nyblade, A.A. and Robinson, S.W., 1994, The African superswell. *Geophysical Research Letters*, **42**, pp. 765–768.

Partridge, T.C. and Maud, R.R., 1987, Geomorphic evolution of southern Africa since the Mesozoic. *South African Journal of Geology*, **90**, pp. 179–208.

Reuter, H.I., Nelson, A. and Jarvis, A., 2007, An evaluation of void filling interpolation methods for SRTM data. *International Journal of Geographic Information Science*, **21**(9), pp. 983–1008.

Robert, M., 1946, *Le Congo Physique*, 3rd edition, (Liège: Vaillant-Carmanne), p. 499.

Roberts, E.M., Stevens, N.J., O'Connor, P.M., Dirks, P.H.G.M., Gottfried, M.D., Clydge, W.C., Armstrong, R.A., Kemp, A.I.S. and Hemming, S., 2012, Initiation of the western branch of the Eastern African Rift coeval with the eastern branch. *Nature Geoscience* **5**(4), pp. 289–294.

Rodriguez, E., Morris, C.S. and Belz, J.E., 2006, A global assessment of the SRTM performance. *Photogrammetric Engineering and Remote Sensing*, **72**(3), pp. 249–260.

Runge, J., 2007, The Congo River, Central Africa. In *Large Rivers: Geomorphology and Management*, edited by Gupta, A., (New York: Wiley), pp. 293–308.

Schumm, S.A., 1977, *The fluvial system*, (New York: Wiley), pp. 338.

Schumm, S.A., 2005, *River variability and complexity*, (Cambridge: Cambridge University Press).

Sinha, S.K. and Parker, G., 1996, Causes of concavity in longitudinal profiles of rivers. *Water Resources Research*, **32**(5), pp. 1417–1428.

Steel, E.A., 1917, Zambezi-Congo Watershed. *Geographical Journal*, pp. 180–193.

Tarboton, D.G., Bras, R.L. and Rodrígues-Iturbe, I., 1988, The fractal nature of river networks. *Water Resources Research*, **24**(8), pp. 1317–1322.

Tooth, S., McCarthy, T.S., Brandt, D., Hancox, P.J. and Morris, R., 2002, Geological controls on the formation of alluvial meanders and floodplain wetlands: the example of the Klip River, eastern Free State, South Africa. *Earth Surface Processes and Landforms*, **27**, pp. 797–815.

Torsvik, T.H. and Cocks, L.R.M., 2009, The Lower Palaeozoic palaeogeographical evolution of the northeastern and eastern peri-Gondwanan margin from Turkey to New Zealand. In *Early Palaeozoic Peri-Gondwana Terranes: New Insights from Tectonics and Biogeography*, edited by Bassett, M.G., (London: Geological Society), Special Publication, **325**, pp. 3–21.

Tucker, J.C., Grant, D.M. and Dykstra, J.D., 2004, NASA's Global Orthorectified Landsat Data Set. *Photogrammetric Engineering and Remote Sensing*, **70**(3), pp. 313–322.

Turcotte, D.L., 1992, *Fractal and chaos in geology and geophysics*, (Cambridge: Cambridge University Press), pp. 221.

US Geological Survey, n.d., [online] Available at: <http://edc.usgs.gov/products/elevation/srtmdted.html>. [Accessed 2009].

Veatch, A.C., 1935, Evolution of the Congo Basin. *Memoir Geological Society of America*, **3**, pp. 183.

Wellington, J. H., 1949, A new development scheme for the Okovango Delta, Northern Kalahari. *Geographical Journal*, pp. 62–69.

Wellington, J.H., 1955, *Southern Africa—a Geographic Study, Volume 1, Physical Geography*, (Cambridge: Cambridge University Press), pp. 528.

Whipple, K.X., 2004, Bedrock rivers and the geomorphology of active orogens. *Annual Review Earth Planetary Sciences*, **32**, pp. 151–185.

Wichura, H., Bousquet, R., Oberhänsli, R., Strecker, M.R. and Trauth, M.H., 2011, The Mid-Miocene East African Plateau: A pre-rift topographic model inferred from the emplacement of the phonolitic Yatta lava flow, Kenya. *Geological Society, Special Publications*, **357**(1), pp. 285–300.

CHAPTER 4

Palynological evidence (*Apectodinium*) of the Paleocene–Eocene Thermal Maximum (PETM) event in the sediments of the Oshosun Formation, Eastern Dahomey Basin, Southwest Nigeria

Samson Bankole
Department of Geosciences, University of Lagos, Nigeria

Eckert Schrank & Bernd-D. Erdtmann
Institut für Angewandte Geowissenschaften, Technische Universität Berlin, Germany

Samuel Akande & Olabisi Adekeye
Department of Geology and Mineral Science, University of Ilorin, Nigeria

Samuel Olobaniyi & Olusola Dublin-Green
Department of Geosciences, University of Lagos, Nigeria

Akintunde Akintola
Department of Geology, Olabisi Onabanjo University, Ago-Iwoye, Nigeria

ABSTRACT: Subsurface shale samples from the Late Paleocene-Early Eocene Oshosun Formation recovered from IB10 Borehole in the Dahomey Basin were subjected to palynological investigation. The palynomorph specimens recovered are moderately well preserved. The investigation revealed the abundance of the thermophilic form, *Apectodinium* in some of the samples. Drastic abundance of *Apectodinium* coincides with a global hyperthermal event, the Paleocene-Eocene Thermal Maximum (PETM). The assemblages of the investigated unit are generally dominated by dinoflagellate cysts represented by *Homotryblium*, *Senegalinium*, *Phelodinium* and *Kallosphaeridium*. The evidence of PETM event is provided with the dramatic increase in abundance of the genus *Apectodinium* at an interval in the investigated borehole sediments. The abundance, distribution pattern and morphological diversification of *Apectodinium* in the Oshosun Formation correlate well with results from other PETM sites around world.

4.1 INTRODUCTION

The Paleocene-Eocene Thermal Maximum (PETM) is a global warming event accompanied by associated environmental and geochemical changes that have been documented in marine sediments across the globe (Zachos *et al.*, 2001; Crouch *et al.*, 2001;

Sluijs *et al.*, 2007; Sluijs *et al.*, 2014; Awad and Ikuenobe-Oboh, 2016). This event which was first recognized in the Antarctic waters by Kenneth and Stott (1991) was dated to be circa 55 ma (Norris and Röhl, 1999) and it is reported to represent a geologically brief time interval of approximately 220 ka (Crouch *et al.*, 2001; Sluijs *et al.*, 2014). The ocean-atmosphere system was apparently affected by a temperature increase of between 4–8°C (Kennett and Stott, 1991). During the period, the ocean-atmosphere was also reported to witness enormous injection of ^{13}C-depleted carbon (Sluijs *et al.*, 2014; Crouch *et al.*, 2001). The consequence of the injection is the negative carbon isotope excursion (CIE). The injection of ^{13}C is attributed to rapid release and oxidation of methane from marine sediments (Gibbs *et al.*, 2006). One of the significant biotic responses to the PETM is the increase in abundance and diversification of the dinoflagellate genus *Apectodinium* (Sluijs *et al.*, 2007; Bujak and Brinkhuis, 1998). This global increase in abundance in *Apectodinium* have been attributed to global increase in temperature during the Paleocene-Eocene transition. Other global events associated with the PETM include acidification of the ocean, terrestrial mammal migration and diversification, global sea level rise (Sluijs *et al.*, 2007), and prominent acceleration in origination and extinction rates of calcareous nannoplankton (Gibbs, 2006).

Previous works which have reported the presence of *Apectodinium* in Nigerian basins include Jan Du Chêne and Adediran (1985), Bankole *et al.*, (2006) and 2007 in the Dahomey Basin, and Oloto (1992) in the Niger Delta. The listed works were merely on taxonomy. Also, recent work by Awad and Oboh-Ikuenobe (2016) in the Côte d'Ivoire-Ghana Transform Margin of West Africa reported *Apectodinium* acme linked to Paleocene-Eocene Thermal Maximum (PETM). The focus of the present study is to report the abundance and distribution of *Apectodinium* in the sediments of the Late Paleocene-Early Eocene (Bankole *et al.*, 2006) Oshosun Formation and link it to PETM event.

4.2 GEOLOGICAL SETTING

The Dahomey (Benin) Basin is one of the African coastal basins, extending from the south-eastern part of Ghana through Togo and the Benin Republic to the south-western Nigeria, where it terminates at the Okitipupa Ridge. The evolution of the basin started in the Maastrichtian (Adediran and Adegoke, 1987). The eastern sector of the basin (Figure 1) is entirely located in Nigeria. The Cretaceous stratigraphy of the basin is represented by the Ise, Afowo, and the Araromi Formations (Figure 2) altogether constituting the Abeokuta Group (Omatsola and Adegoke, 1981). This stratigraphic interval was reviewed by Okosun (1990) and considered to consist of fine- to coarse-grained, unconsolidated sand with shale, mud, and limestone intercalations.

The submergence of the basin in the Paleocene led to the deposition of the Ewekoro Formation, which consists of several limestone varieties (sandy biomicrosparite, shelly biomicrite, and algal biosparite) at its type locality in Ewekoro and also at the Ibese and Sagamu Quarries. The extensive limestone sequence is overlain by a thin (10–20 cm) glauconitic sand bed, separating the limestone from the overlying Oshosun Formation.

The Oshosun Formation is considered to be deposited during the Late Paleocene–Early Eocene (Jan Du Chêne and Adediran, 1985; Bankole *et al.*, 2006). It is a massive, grey to dark coloured and laminated shale sequence (Figure 3). Capping the sequence are the brown to red coloured, poorly sorted, and poorly consolidated sandstones of the Ilaro Formation exposed along Papalanto-Ilaro road and some other localities in the basin.

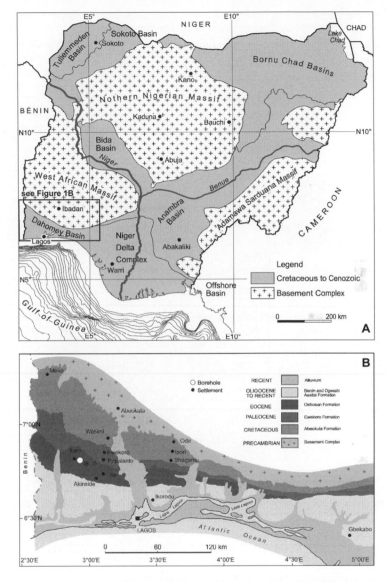

Figure 1. A. Geological sketch map of Nigeria (modified after Whiteman, 1982), Box shows the location of the Dahomey Basin. **B.** Geological map of eastern Dahomey Basin (modified after Billman, 1976).

4.3 MATERIAL AND METHOD

Core samples from a borehole (IB10: Figure 3) were subjected to standard palynological analysis. Approximately 10 g of each sample were crushed to approximately 2 mm size before the treatment with 10% hydrochloric acid (HCl) to remove the carbonates that might be present in the sample. To ensure a complete removal of the carbonates, the HCl was added in excess until no more effervescence was given.

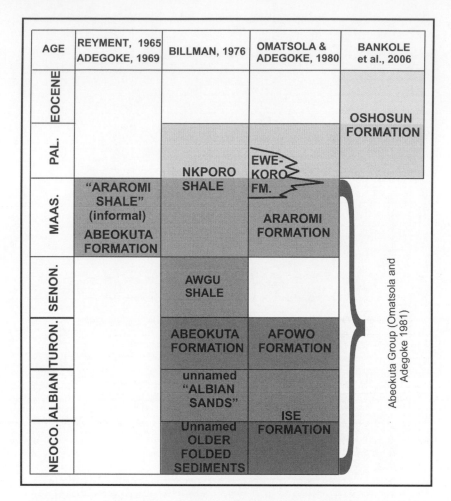

Figure 2. Age and the stratigraphic relationship of the formations of the Dahomey Basin
(Bankole *et al.*, 2006).

The process was followed by the treatment of the samples with HF (40% concentration) for the removal of the silicates. The content was allowed to settle for 24 hours before the HF was carefully decanted. Thereafter, complete neutralisation was carried out with distilled water. To remove fluoro-silicate compounds that might be formed from the reaction of HF, the residue was again treated with HCl. This was followed by another round of neutralisation with distilled water. No oxidation step was involved in this preparation as the process could selectively destroy the dinoflagellates cysts (dinocysts). The whole process was carried out in a fume-chamber.

The residue was then sieved using 15 μm sieve for the recovery of the paly-nomorphs. One to two drops of the > 15 μm residue was mounted on microscope glass slides using a glycerine jelly medium. For each sample, two slides were prepared. Scanning for visual assessment was carried out with the aid of a Leitz Laborlux S microscope. Pictures were taking using a Wild Leitz MPS52 camera. All slides with sieved and unsieved residue are stored at the Institut für Angewandte Geowissenschaften, Technical University Berlin.

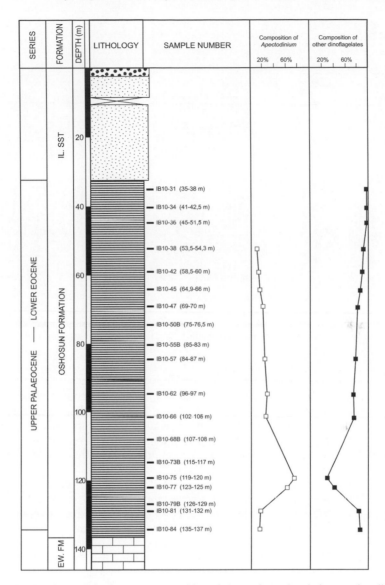

Figure 3. Abundance and percentage composition of *Apectodinium* in relation to other dinoflagellate group and pollen/spores in the IB10 section. B = Barren.

4.4 RESULTS

Nineteen samples were analysed. Fourteen yielded palynomorphs, while the remaining five samples were either completely barren or extremely poor. A minimum of two hundred palynomorph specimens were counted per sample. Generally, the assemblage is dominated by dinoflagellate cysts, constituting ~85% of the total palynomorphs recovered. IB10-84 (Figure 3) is the most prolific of all the samples. Dominant dinocyst genera in IB10-84 sample include *Homotryblium* and *Senegalinium*. Other common dinoflagellates occurrence in this sample are *Apectodinium homomorphum* (Plate 1),

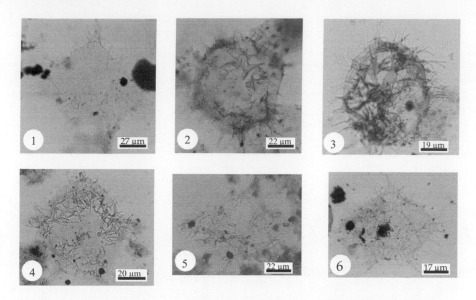

Plate 1. Some specimens of *Apectodinium* from IB10 Borehole section: Figures 1–6: *Apectodinium quinquelatum*, Figures 2–4: *Apectodinium homomorphum*, Figure 5: *Apectodinium* sp. A.

Phelodinium magnificum, and species with close affinity to the genus *Kallosphaeridium.* A few meters up section (Samples IB10-77 and IB10-75), there was a relative sharp increase in palynofloral abundance and diversity, most especially in Sample IB10-77. Concomitant to this is the increase in abundance and diversity of the *Apectodinium* genus with accompanying decrease in the abundance of other dinocyst group. The abundance of *Apectodinium* is about 60% of the total dinocyst assemblage in sample IB10-77 and increased to about 75% in IB10-75. The *Apectodinium* species include *Apectodinium homomorphum, A. quinquelatum, A.* sp., and *A.* sp. 1.

There was a drastic reduction in *Apectodinium* abundance and diversity from sample IB10-66, while the abundance of the other dinocyst group increased. The increase in percentage contribution of land-derived palynoflora observed from IB10-75 persisted in IB10-66 and IB10-62, where it reaches 17%. The palynomorphs in Sample IB10-66 consists of 67% dinoflagellate cysts, 17% pollen/spores, 6% inaperturates, and 6% chitinous micro-foraminifera lining. No drastic change in assemblage contribution from IB10-62 through IB10-31 was observed. However of note is that no single *Apectodinium* is recorded from IB10-36 through IB10-31.

4.5 DISCUSSION

It is evident from the studied section that the palynomorphs assemblages recorded were largely dominated by the dinoflagellate cysts. Of special note is the *Apectodinium* acme recorded in samples IB10-77 and IB10-75, which is linked to the PETM. Concomitant with this spike is the relative increase in the abundance of pollen/spores and the drastic reduction in the occurrence of other dinoflagellates forms at this interval. Increased abundance of pollen and spore at the interval could be the consequence of encroachment of continental areas during the global sea level rise associated with PETM. An indication of negative response to extreme temperature by the other group of dinoflagellates is suggested by their drastically reduced abundance at the interval. The onset of the *Apectodinium* acme has been reported (Crouch *et al.*, 2001) to have rapidly occurred globally at the beginning of PETM. The cause of this increase in abundance of *Apectodinium* is attributed to increased seawater temperatures during the Latest Paleocene and subsequent increased productivity in marginal seas at the beginning of PETM (Crouch *et al.*, 2001; Thomas, 1998). The anomalous temperature increase (between 4° to 8°C) witnessed during the Late Paleocene–Early Eocene interval has been linked (in one of the leading hypothesis) to the dissociation of 1400 to 2800 gigatonnes of methane from ocean clathrates, which resulted in a large negative carbon isotope excursion and huge carbonate dissolution in marine sediments (Lourens *et al.*, 2005 and the references therein). A drastic reduction in *Apectodinium* genus with corresponding increased abundance of the other dinoflagellate group characterized samples IB10-66 through IB10-38. This event is suggestive of a relative reduction in temperature, which might have favoured the productivity of the other dinoflagellate group.

4.6 CONCLUSIONS

Palynonological evidence of PETM event in the eastern sector of the Dahomey Basin is presented through the distribution pattern and mode of occurrence of the dinoflagellate thermophilic genus, *Apectodinium*. The *Apectodinium* acme observed in samples IB10-77 and IB10-75 in the investigated section has been carefully compared with result of similar assemblages where PETM have been directly linked with *Apectodinium* acme around the world. Increased abundance of pollen and spores is indicative of improved supply of continentally derived material to the marine environment as a result of the encroachment of the continent during sea level rise.

ACKNOWLEDGEMENTS

The investigation was conducted within the framework of VW-Foundation (Hannover, Germany) funded research work in the Dahomey Basin, Nigeria, and the Potiguar Basin, Brasil (Grant AZ I/77620). The authors, therefore, acknowledge that the funding for field studies and laboratory analyses was provided through the grant. Thanks are due to the two anonymous reviewers for their constructive criticisms that have hugely improved the quality of the paper.

REFERENCES

Adegoke, O.S., 1969, Eocene stratigraphy of southern Nigeria. *Bur. Rech. Geol. Min. Mem.*, **69**, pp. 23–48.

Adediran, S.A. and Adegoke, O.S., 1987, Evolution of the sedimentary basins of the Gulf of Guinea. In *Current Research in African Earth Science*, edited by Matheis, G. and Schandelmeier, H., (Rotterdam: Balkema), pp. 283–286.

Awad, W.K. and Oboh-Ikuenobe, F.E., 2016, Early Paleogene dinoflagellate cysts from ODP Hole 959D, Côte d'Ivoire-Ghana Transform Margin, West Africa: New Species, biostratigraphy and paleoenvironmental implications. *Journal of African Earth Sciences*, **123**, pp. 123–144.

Bankole, S.I., Schrank, E. and Erdtmann, B.-D., 2007, Palynology of the Paleogene Oshosun Formation in the Dahomey Basin, Southwestern Nigeria. *Revista Espanola de Micropaleontologia*, **39**(1–2), pp. 29–44.

Bankole, S.I., Schrank, E., Erdtmann, B.-D. and Akande, S.O., 2006, Palynostratigraphic age and paleoenvironments of the newly exposed section of the Oshosun Formation in the Sagamu Quarry, Dahomey Basin, South-Western Nigeria. Nigeria *Association of Petroleum Explorationists Bull.*, **19**(1), pp. 25–34.

Billman, H.G., 1976, Offshore stratigraphy and paleontology of the Dahomey Embayment. *Proc. 7th African Micropaleontology Coll.*, Ile-Ife, pp. 27–42.

Bujak, J.P. and Brinkhuis, H., 1998, Global warming and dinocyst changes across the Paleocene/Eocene Epoch boundary. In *Late Paleocene–Early Eocene biotic and climatic events in the marine terrestrial records*, edited by Aubry, M.-P., Lucas, S. and Berggren, W., (New York: Columbia University Press), pp. 277–295.

Crouch, E.M., Heilmann-Clausen, C., Brinkhuis, H., Morgans, H.E.G., Rogers, K.M., Egger, H. and Schmitz, B., 2001, Global dinoflagellate event associated with the Late Paleocene Thermal Maximum. *Geology*, **29**(4), pp. 315–318.

Gibbs, S.J., Bown, P.R., Sessa, J.A., Bralower, T.J. and Wilson, P.A., 2006, Nannoplankton extinction and origination across the Paleocene-Eocene Thermal Maximum. *Science*, **314**, pp. 1770–1773.

Jan du Chêne, R.E. and Adediran, S.A., 1985, Late Paleocene to Early Eocene dinoflagellates from Nigeria. *Cahiers Micropaléontologie*, **3**, pp. 1–38.

Kennett, J.P. and Stott, L.D., 1991, Abrupt deep sea warming, paleoceanographic changes and benthic extinction at the end of the Paleocene. *Nature*, **353**, pp. 225–229.

Lourens, L.J., Sluijs, A., Kroon, D., Zachos, J.C., Thomas, E., Röhl, U., Bowles, J. and Raffi, I., 2005, Astronomical pacing of Late Palaeocene to Early Eocene global warming events. *Nature*, **435**(7045), pp. 1083–1087.

Norris, R.D. and Röhl, U., 1999, Carbon cycling and chronology of climate warming during the Paleocene/Eocene transition. *Nature*, **401**, pp. 775–778.

Okosun, E.A., 1990, A review of the Cretaceous stratigraphy of the Dahomey Embayment, West Africa. *Cretaceous Research*, **11**, pp. 17–27.

Oloto, I.N., 1992, Succession of palynomorphs from the early Eocene of Gbekebo-1 well in S.W. Nigeria. *Journal of African Earth Sciences*, **15**, pp. 441–452.

Omatsola, M.E. and Adegoke. O S., 1981, Tectonic evolution and cretaceous stratigraphy of the Dahomey Basin. *Journal of Mining and Geol.*, **18**, pp. 130–137.

Reyment, R.A., 1965, Aspect of the Geology of Nigeria, (Ibadan: Ibadan University Press), **145**.

Sluijs, A., Van Roij, L., Harrington, G.J., Schouten, S., Sessa, J.A., LeVay, L.J., Reichart, G.J. and Slomp, C.P., 2014, Warming, euxinia and sea level rise during the Paleocene–Eocene Thermal Maximum on the gulf coastal plain: Implications for ocean and oxygenation and nutrient cycling. *Climate of the Past*, **10**, pp. 1421–1439.

Sluijs, A., Bowen, G.J., Brinkhuis, H., Lourens, L.J. and Thomas, E., 2007, The Paleocene–Eocene Thermal Maximum upper greenhouse: biotic and geochemical signatures, age models and mechanisms of global change. In *Deep time perspective on climate change: Marrying the signal from computer models and biological proxies*, edited by Williams, M., Haywood, A.M., Gregory, F.J. and Schmidt, D.N., (London: The Geological Society), The micropalaeontological Society, Special Publications, pp. 323–347.

Thomas, E., 1998, Biogeography of the Late Paleocene benthic foraminiferal extinction. In *Late Paleocene-Early Eocene biotic and climatic events in the marine terrestrial records*, edited by Aubry, M.-P., Lucas, S. and Berggren, W., (New York: Columbia University Press), pp. 214–243.

Zachos, Z., Pagani, M., Sloan, L., Thomas, E. and Billups, K., 2001, Trends, rhythms, and aberrations in global climate 65 Ma to present. *Science*, **292**, pp. 686–693.

CHAPTER 5

Sedimentological, palynological and stable isotopes studies on Quaternary to Neogene sediments of the eastern Dahomey Basin, Lagos, Nigeria

Olugbenga A. Boboye & Dupe Egbeola
Department of Geology, University of Ibadan, Ibadan, Nigeria

ABSTRACT: Sediments out of three boreholes located in the Lagos lagoon area (wells at Victoria Island, Apapa, and Lekki) were studied for provenance, environment of deposition, maturity, and age of samples. It was also focused on the palaeoclimatic conditions (ambient water temperature) that prevailed during deposition. Samples were medium-coarse grained, moderately to very poorly sorted, strongly positive skewed to near symetrical, and very platykurtic; and they show extremely leptokurtic trends. This is indicative of a river environment that deposited fluvial sediments, suggesting a high to medium energy transporting medium. The heavy mineral assemblages were characterized by zircon, tourmaline, rutile, garnet, epidote, staurolite, apatite, and sillimanite, suggesting igneous and metamorphic provenance. The average Zircon-Tourmaline-Rutile (ZTR) index was 53.9%, indicating texturally immature sediments. Quartz is the most abundant mineral accounting for 90% to 95%. The Quartz-Feldspar-Rock fragments (QFR) ternary plots revealed predominantly super mature quartz arenites. Palynomorphs revealed abundance of mangrove swamp forest species in association with freshwater swamp forest species and freshwater with algae (*Botryococcus* sp.). This suggests a mangrove environment with a high influx of freshwater. The sediments were inferred to be of Quaternary and Neogene age. Pebble morphometric studies revealed a littoral setting, and the plot of maximum projection sphericity against oblate prolate index again suggested fluvial origin of sediments. The pebble forms were dominantly compact and compact-bladed, which further supported a process of fluvial dynamics. Oxygen and carbon isotopic composition of bivalves from Lekki well ranged from −4.93‰ to −2.94‰ VPDB (mean value of −4.22‰) and −10.35‰ to −3.71‰ VPDB (mean value of −7.43‰); while the Apapa well shell fragments (*Venericardia planicosta*) revealed $\delta^{18}O$ values of −2.48‰ and −2.19‰ VPDB, $\delta^{13}C$ values of −3.27‰ and −1.96‰ VPDB, respectively. The results showed a positive correlation between oxygen and carbon isotopic compositions, which suggests a seasonal climate change that might have influenced carbon sources to some degree. The estimated ambient water palaeo-temperatures during the Neogene might have varied in the range between 28.93°C to 40.82°C.

5.1 INTRODUCTION

Series of research work had been conducted on the Dahomey Basin. Bassey (1996) reported that the Nigerian coastal region is vested with numerous laden rivers that flow more or less southward through an extensive hinterland into adjourning Atlantic Ocean. The sediments so supplied are dispersed by tides generated by longshore drift that flows directly onto the coast. Bassey (1996) studied the grain sizes and chemical composition of beach sands on Victoria Island and reported that the main sediment source is suggested to be of highly metamorphosed and acidic igneous rocks of the adjoining Nigerian basement complex, though some were transported by longshore currents from Ghana-Togo coast to the Lagos shore. The Dahomey Basin is one of the sedimentary basins on the continental margin of the Gulf of Guinea, extending from south-eastern Ghana in the west to the western flank of the Niger Delta (Jones and Hockey, 1964; Omatsola and Adegoke, 1981). It is bounded on the West by the Ghana ridge, the extension of the Romanche Fracture Zones; while in the east, it is bounded by the Benin Hinge line (Figure 1a). The origin is closely related to the rifting and separation of the African and South American plates during the late Jurassic and early Cretaceous (Brownfield and Charpentier, 2006; Billman, 1976; Adegoke, 1969, 1977; Elvsborg and Dalode, 1985; Omatsola and Adegoke, 1981; Ogbe, 1972; Ako et al., 1981). The sequence of rock unit from top are the Coastal Plain Sands (Oligocene to recent), Ilaro Formation (Eocene), Oshosun Formation (Eocene), Akinbo Formation (Paleocene), Ewekoro Formation (Palaeocene), and Abeokuta Group (Maastrichtian to Paleocene) (Jones and Hockey, 1964). The climate change has resulted in repeated environmental changes world-wide. Information on the dynamics of these changes is often preserved in sedimentary deposits associated with the margins of marine, lacustrine, or glacial environments. However, most of these palaeoenvironmental records are limited in the degree of age control (Molodkov and Bolikhovskaya, 2002). Amongst other West African researchers are Salzmann et al. (2002), Boboye and Nwosu (2013), and Sowunmi (1981, 2004).

This study aims at investigating the effect of climatic variability during the Neogene to Quaternary using faunal and flora remains and the description of the stratigraphical and geological context to determine the lithofacies and deduce the provenance, transportation history, environment of deposition, as well as age and climatic conditions that prevailed during deposition.

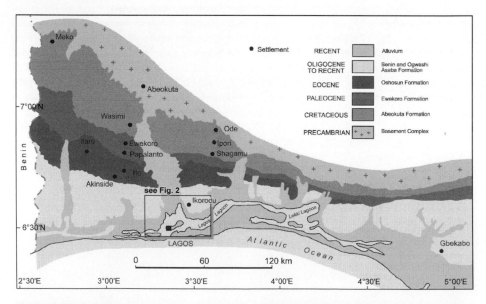

Figure 1a. Geological map of the Dahomey Basin (modified after Billman, 1976). Formations covering Precambrian basement rocks (from top base): Alluvium (recent), Coastal Plain Sands (recent to Oligocene); Ilaro and Oshosun (Eocene); Ewekoro (Paleocene); Abeokuta Group (Maastrichtian to Paleocene).

5.2 GEOLOGICAL SETTING AND STRATIGRAPHY OF THE STUDY AREA

The Dahomey Basin is a marginal pull-apart basin (Klemme, 1975) or margin sag basin (Kingston *et al.*, 1983a, 1983b), which was initiated during the early Cretaceous separation of Africa and South American lithospheric plates (Figures 1a, 1b). Geology and stratigraphy of the Dahomey Basin has been described by Jones and Hockey (1964), Elueze and Nton (2004), Omatsola and Adegoke (1981), Akinmosin and Osinowo (2008), and Boboye and Nwosu (2013). In most part of the Basin, the stratigraphy is dominated by sand and shale alternations with minor proportion of limestone (Agagu, 1985). Eight lithostratigraphic units have been identified and described. Stratigraphic studies of Dahomey Basin were reported by various workers amongst which are Jones and Hockey (1964), who studied outcrop units of crystalline

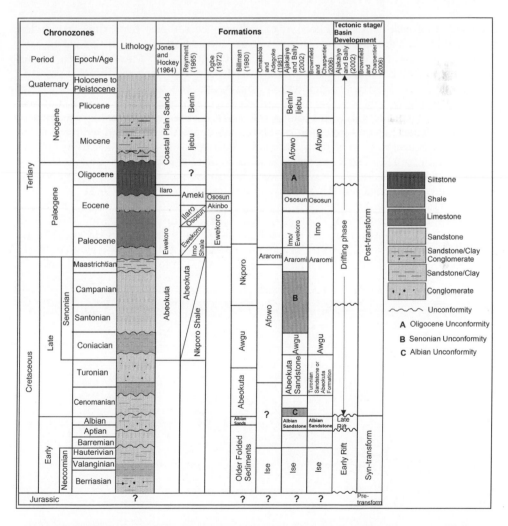

Figure 1b. Generalized stratigraphy showing age, lithology, and sequence of the formations; and tectonic stage of basin development for the Dahomey (Benin) Basin in Nigeria (Olabode and Mohammed, 2016, p. 213).

basement complex of southwestern Nigeria. Reyment (1965) classified the Dahomey Basin as one of the components of the Nigerian basin of which the Dahomey sector is the left extension. Later studies proposed other stratigraphies of the basin-onshore and offshore sediments (Antolini, 1968; Adegoke, 1969; Fayose, 1970; Ogbe, 1972; Ako *et al.*, 1981; Omatsola and Adegoke, 1981, and Okosun, 1987).

The general sequences of the rock units from the youngest are the Coastal Plain Sands (Oligocene to Recent), Ilaro Formation (Eocene), Oshosun Formation (Eocene), Akinbo Formation (Paleocene), Ewekoro Formation (Paleocene), and Abeokuta Group consisting of three formations, namely, Ise (Valanginian to Barremian), Afowo (Maastrichtian), and Araromi formations (Maastrichtian to Paleocene) (Figure 1b).

5.3 MATERIAL AND METHODS

The field approach entailed collection of sub-surface samples from three boreholes (wells) drilled in the study area. These wells are located at Apapa (N 06° 26′ 17.58″, E 03° 22′ 57.89″), Lekki (N 06° 27′ 32.35″, E 03° 0.36′ 5.48″), and Victoria Island (N 06° 25′ 51.59″, E 03° 26′ 2.88″) (Figure 2). After stratigraphy and lithologic description samples were retrieved at 3 m intervals from Apapa well with a total depth of 228 m (Figure 3) and Lekki well with 246 m total depth (Figure 4); while the Victoria Island well was sampled at 6 m intervals with a total depth of 204 m (Figure 5). Sampling was carried out aided by hand lens and stereo-binocular microscope with a standard sedimentological chart to delineate lithologic units. Samples were described in terms of texture, colour, grain size, and composition. Generally, the wells comprised of siltstone, sandstone, claystone ranging from medium to very coarse grained, moderately well sorted to very poorly sorted, and pebbly conglomerates with fossil shell fragments (Figures 3 to 5).

The samples were systematically selected for granulometry at depths of 9 m, 21 m, 42 m, 57 m, 69 m, 78 m, 96 m, 123 m, 132 m, 144 m, 162 m, 183 m, 195 m, 207 m, and 219 m for Apapa well; at 18 m, 27 m, 63 m, 75 m, 87 m, 99 m, 111 m, 126 m, 138 m, 150 m, 156 m, 171 m, 195 m, 207 m, 219 m, and 231 m for Lekki well; and at 36 m, 66 m, 78 m, 96 m, 102 m, 108 m, 144 m, 162 m, 186 m, and 204 m for Victoria Island well.

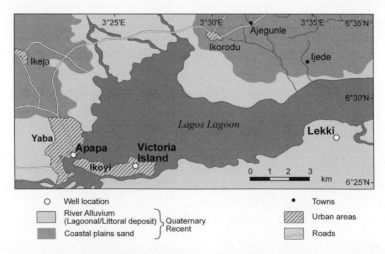

Figure 2. Geological sketch map of the Lagos lagoon with location of boreholes/wells sites at Apapa, Lekki, and Victoria Island (Figures 3–5).

Subsequently, they were air-dried, disaggregated and passed through varying sieve sizes of 4.75 millimetres (mm), 3.35 mm, 2.36 mm, 1.70 mm, 1.18 mm, 850 µm, 600 µm, 425 µm, 300 µm, 212 µm, 150 µm, 106 µm, and 75 µm.

A total of fifteen samples were used for heavy mineral separation—four samples from Apapa well (AP/W/12 m, AP/W/117 m, AP/W/153 m, AP/W/171 m, AP/W/174 m); five samples from Lekki well (LK/W/9 m, LK/W/18 m, LK/W/39 m, LK/W/78 m, LK/W/153 m); and six samples from Victoria Island well (VI/W/6 m, VI/W/12 m, VI/W/18 m, VI/W/72 m, VI/W/114 m, VI/W/153 m).

Apparatus and reagents employed include bromoform, acetone, dilute HCl, Canada Balsam, filter paper, funnels, and glass slides. Each sample was sieved and

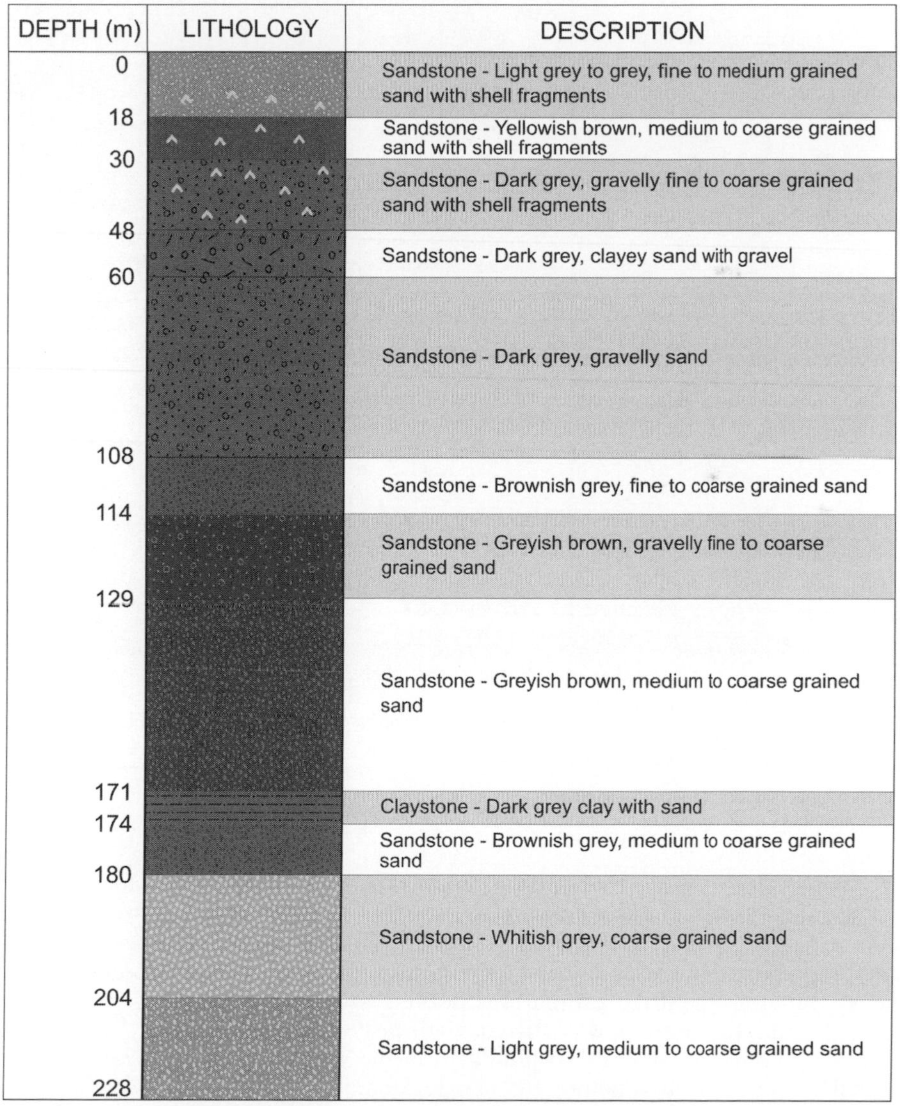

DEPTH (m)	LITHOLOGY	DESCRIPTION
0		Sandstone - Light grey to grey, fine to medium grained sand with shell fragments
18		Sandstone - Yellowish brown, medium to coarse grained sand with shell fragments
30		Sandstone - Dark grey, gravelly fine to coarse grained sand with shell fragments
48		
60		Sandstone - Dark grey, clayey sand with gravel
		Sandstone - Dark grey, gravelly sand
108		
114		Sandstone - Brownish grey, fine to coarse grained sand
129		Sandstone - Greyish brown, gravelly fine to coarse grained sand
		Sandstone - Greyish brown, medium to coarse grained sand
171		Claystone - Dark grey clay with sand
174		Sandstone - Brownish grey, medium to coarse grained sand
180		
		Sandstone - Whitish grey, coarse grained sand
204		
		Sandstone - Light grey, medium to coarse grained sand
228		

Figure 3. Stratigraphy and lithology of the Apapa well.

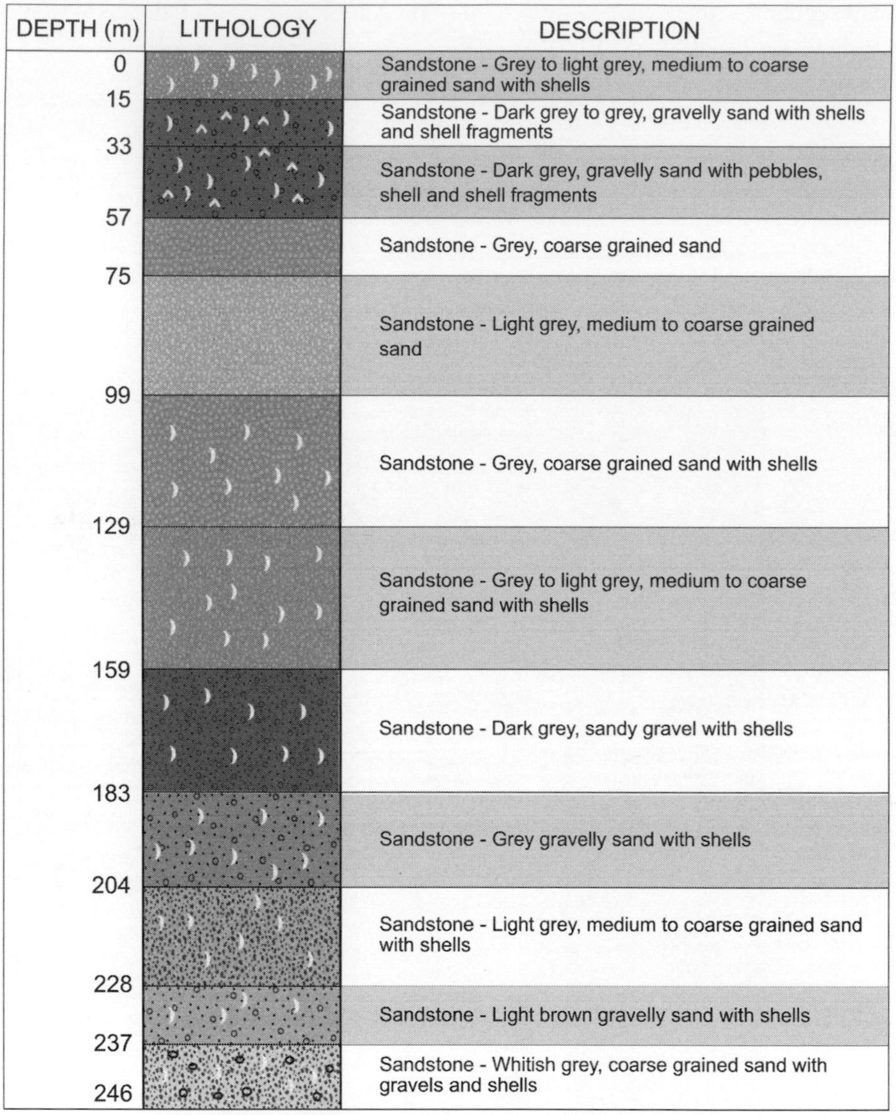

DEPTH (m)	LITHOLOGY	DESCRIPTION
0		Sandstone - Grey to light grey, medium to coarse grained sand with shells
15		
33		Sandstone - Dark grey to grey, gravelly sand with shells and shell fragments
		Sandstone - Dark grey, gravelly sand with pebbles, shell and shell fragments
57		
75		Sandstone - Grey, coarse grained sand
		Sandstone - Light grey, medium to coarse grained sand
99		
		Sandstone - Grey, coarse grained sand with shells
129		
		Sandstone - Grey to light grey, medium to coarse grained sand with shells
159		
		Sandstone - Dark grey, sandy gravel with shells
183		
		Sandstone - Grey gravelly sand with shells
204		
		Sandstone - Light grey, medium to coarse grained sand with shells
228		
237		Sandstone - Light brown gravelly sand with shells
246		Sandstone - Whitish grey, coarse grained sand with gravels and shells

Figure 4. Stratigraphy and lithology of the Lekki well.

the fraction retained on sieve of 212 μm mesh was used in the heavy mineral separation. The samples were treated with dilute hydrochloric acid for about two minutes to remove carbonates, washed with distilled water to remove acid effect on the samples, and were later dried. Gravity settling techniques was employed for the separation of the heavy mineral. The dried sample was poured into a separating funnel containing bromoform and the content was stirred vigorously and allowed to settle for about 20 minutes. The heavy minerals accumulated in the stem of the funnel were dropped into the filtering funnel by opening the stop lock carefully to drain the heavy minerals. The heavy minerals were then washed with acetone, allowed to dry, and mounted on glass slides with Canada balsam.

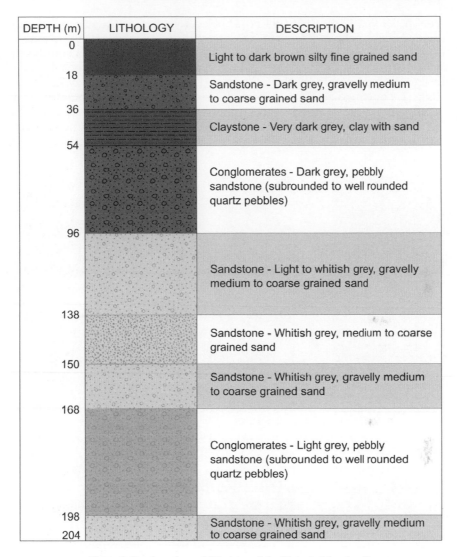

DEPTH (m)	LITHOLOGY	DESCRIPTION
0		Light to dark brown silty fine grained sand
18		
		Sandstone - Dark grey, gravelly medium to coarse grained sand
36		
		Claystone - Very dark grey, clay with sand
54		
		Conglomerates - Dark grey, pebbly sandstone (subrounded to well rounded quartz pebbles)
96		
		Sandstone - Light to whitish grey, gravelly medium to coarse grained sand
138		
		Sandstone - Whitish grey, medium to coarse grained sand
150		
		Sandstone - Whitish grey, gravelly medium to coarse grained sand
168		
		Conglomerates - Light grey, pebbly sandstone (subrounded to well rounded quartz pebbles)
198		Sandstone - Whitish grey, gravelly medium to coarse grained sand
204		

Figure 5. Stratigraphy and lithology of the Victoria Island well.

Petrographic analysis was carried out on a total of fifteen ditch samples—six samples from Apapa well (AP/W/12 m, AP/W/117 m, AP/W/153 m, AP/W/171 m, AP/W/174 m, AP/W/222 m); five samples from Lekki well (LK/W/9 m, LK/W/18 m, LK/W/39 m, LK/W/78 m, LK/W/153 m); and four samples from Victoria Island well (VI/W/6 m, VI/W/18 m, VI/W/72 m, VI/W/114 m).

They were impregnated with epoxyl and left to core for 24 hours after which the samples were cut to fit on glass slide using the cut-off saw. The surface of the sample was lapped on a glass plate using water and silicon carbide 600 grit (Carborundum).

Pebble morphometry was only done on twelve pebbly conglomerates recovered in the sand units from Victoria Island well (VI/W/66 m, VI/W/72 m, VI/W/78 m, VI/W/84 m, VI/W/90 m, VI/W/96 m, VI/W/114 m, VI/W/174 m, VI/W/180 m,

VI/W/186 m, VI/W/192 m, VI/W/198 m). Pebbles from this well were sub-rounded to well-rounded quartz pebbles (Figure 6). The largest 10 of 12 pebbles (Figure 6) were used for morphometrical analyses. The pebble axes, long (L), intermediate (I) and short (S), were measured with the aid of vernier callipers. The roundness of each pebble was estimated using the image set for pebble roundness (Krumbein, 1941). The values of each of the long (L), intermediate (I), and short (S) axes as well as the roundness for the pebbles in each sample were averaged. The morphometric parameters computed were Coefficient of Flatness, (S/L)*100 (Stratten, 1974; Els, 1988); Elongation Ratio, I/L (Sames, 1966; Lutig, 1962); Flatness Ratio, S/L (Lutig, 1962); Maximum Projection Sphericity

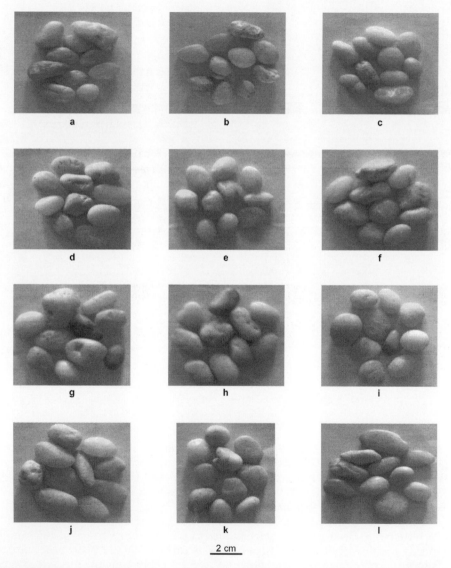

2 cm

Figure 6. Pebbles shapes from Victoria Island well (a) VI/W/066 m, (b) VI/W/072 m, (c) VI/W/078 m, (d) VI/W/084 m, (e) VI/W/090 m, (f) VI/W/096 m, (g) VI/W/114 m, (h) VI/W/174 m, (i) VI/W/180 m, (j) VI/W/186 m (k) VI/W/192 m (l) VI/W/198 m (Scale = 2 cm).

$(S^2/LI)^{1/3}$ (Sneed and Folk, 1958); Oblate—Prolate Index (OPI), $10\{(L–I)/L–S) – 0.50\}/\{S/L\}$ (Dobkins and Folk, 1970); and Form, $(L–I / L–S)$ (Sneed and Folk, 1958).

Palynological observations were based on nine samples only, that were chosen from the respective wells—three samples from Apapa (AP/W/117 m, AP/W/153 m); four from Lekki (LK/W/3 m, LK/W/51 m, LK/W/21 m, LK/W/39 m); and two from Victoria Island (VI/W/42 m, VI/W/54 m).

The samples containing palynomorphs (Table 4) were disaggregated in a mortar and concentrated hydrofluoric acid (HF). HCl (36%) was added, heated in a water bath for 45 minutes, centrifuged, and the supernatant was decanted removing the silico-fluorides. The palynomorphs were separated from the disaggregated silica by adding zinc chloride $(ZnCl_2/HCl)$ to the residue. The samples were mounted using Canada balsam. The slides were studied under a binocular light microscope (Leitz Diaplan) and the taxa identification was done using reference slides, as well as Chevron and Shell (SPDC) albums. The dating was based on STRATCOM Schemes (Evamy *et al.*, 1978; Germeraad *et al.*, 1968). Pollen diagram was plotted using TILIA, while grouping of samples were enhanced by cluster analysis dendrogram by CONISS. Grouping was based on similarities in palynomorphs and general palynomorph assemblage. Palynomorphs were classified into phyto-ecological groups based on the natural habitats of their presumed parent plants (Table 4).

The analysis of stable isotopes $(^{13}C/^{12}C)$ and oxygen $(^{18}O/^{16}O)$ of carbonate of shells and shell fragments was applied to samples recovered from the Apapa and Lekki wells. The shells from Lekki, *Neptunea tabulate,* were composite within specific depth range that contained these shells and also on similarity in sizes, although shell specimens from Apapa well were fragmented (*Venericardia planicosta*) (Figure 7). A total number of six specimens from Lekki well (LK/W/3–30 m (G), LK/W/3–30 m (B), LK/W/33–51 m, LK/W/108–153 m, LK/W/156–198 m, LK/W/207–246 m) and two from Apapa well (AP/W/15–27 m, AP/W/30–36 m) were subjected to oxygen and carbon isotopic analysis (Tables 6 and 7). The specimens were prepared and analysed at Isotech Laboratories,

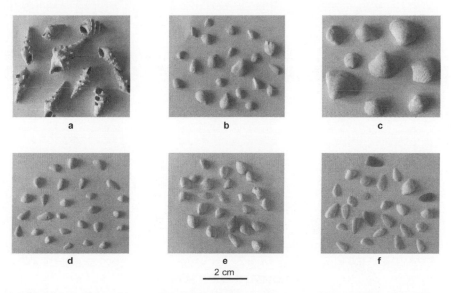

Figure 7. Selected shells for isotopic analysis from Lekki (a) *Neptunea tabulata* (3–30 m), and Apapa well, *Venericardia planicosta* (partly fragmented) recovered at (b) 3–30 m; (c) 33–51 m; (d) 108–153 m; (e) 156–198 m; and (f) 207–246 m.

USA. This was carried out by acid digestion and CF-IRMS (Continuous Flow Isotope Ratio Mass Spectrometer). The GasBench II Isotope Mass Spectrometer (IRMS) system is utilized for the isotopic ratio of carbon ($^{13}C/^{12}C$) and oxygen ($^{18}O/^{16}O$) in carbonate.

The palaeotemperature (t) equation for equilibrium precipitation of carbonates used according to Grossman and Ku (1986) is:

$$t = 20.60 - 4.34\,(c - (w - 0.27)) \qquad (1)$$

where t = temperature (°C); c = $\delta^{18}O$ carbonate (to Vienna-PeeDee Belemnite (VPDB) in ‰), w = $\delta^{18}O$ water (to Vienna-standard mean oceanic water (V-SMOW) in ‰).

5.4 RESULTS AND DISCUSSION

5.4.1 Granulometry

A total number of 15 (Apapa), 17 (Lekki), and 10 (Victoria Island) soft sandstone samples were subjected to granulometric analysis. The mean grain size values (skewness) of the Victoria Island well samples ranged from −0.37 to 0.78 (Figure 8). Two main sediment type sizes were established, that is, coarse sand and very coarse sand, they are obviously associated with high energy conditions. Smaller grains were washed away leaving the coarser grains that were too heavy to be carried away by the current suggesting an upper course of a river regime. The sorting values indicate that all of the

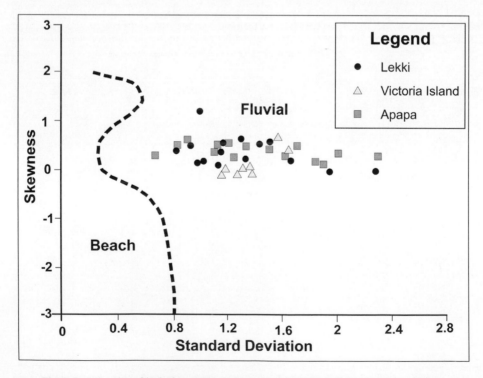

Figure 8. Cross-plot of inclusive graphic skewness and inclusive graphic standard deviation of the studied wells (after Friedman, 1967).

samples that have not been transported far away from their source are poorly sorted. High energy of transportation which did not permit hydraulic sorting was probably associated with flash floods and caused relatively high rates of sedimentation. The skewness value (Figure 8) ranged from −0.06 to 0.71 indicating that 20% of the samples were strongly positive skewed, 20% exceeded the maximum value, and 60% were near symmetrical. The strongly positive skewed nature of the sediments reflects input from various sources of tributaries. The poorly sorted nature of the sediments and the positively skewed nature support a fluvial origin. The graphic kurtosis ranged from 0.51 to 1.09 and indicated that 30.0% were very platykurtic, 20% were platykurtic, and 50% were mesokurtic. This strongly suggests fluvial and/or tidal environment, mainly of river deposits.

5.4.2 Pebble morphometry

The results of pebble morphometric measurements of the pebbly conglomerates recovered from the Victoria Island well are presented in Table 1. It is suggested that pebble suites with mean sphericity of 0.65 and indicated beach processes weakly while those with sphericity values above 0.65 are shaped by fluvial processes (Dobkin and Folk,

Table 1. Mean values of Victoria Island well pebble morphometric data (ten pebbles in each data).

SAMPLE	L (cm)	I (cm)	S (cm)	F.R. (S/L)	E.R. (I/L)	(L-I)/(L-S)	MPSI (S²/LI)^(1/3)	OPI	COEFFICIENT OF FLATNESS (S/L*100) %	FORM NAME	ROUNDNESS %
VI/W/066 m	1.22	1.07	0.87	0.71	0.88	0.43	0.83	−1.00	71	Compact	75
VI/W/072 m	1.15	1.00	0.84	0.73	0.87	0.48	0.85	−0.22	73	Compact	76
VI/W/078 m	1.20	1.02	0.76	0.63	0.85	0.41	0.78	−1.44	63	Compact bladed	84
VI/W/084 m	1.25	1.06	0.87	0.70	0.85	0.50	0.83	0.00	70	Compact bladed	81
VI/W/090 m	1.17	0.99	0.84	0.72	0.85	0.55	0.85	0.63	72	Compact	83
VI/W/096 m	1.15	0.95	0.80	0.70	0.83	0.57	0.84	1.03	70	Compact bladed	79
VI/W/114 m	1.20	1.01	0.91	0.76	0.84	0.66	0.88	2.05	76	Compact	81
VI/W/174 m	1.32	1.09	0.86	0.65	0.83	0.50	0.80	0.00	65	Compact bladed	81
VI/W/180 m	1.17	0.99	0.86	0.74	0.85	0.58	0.86	1.10	74	Compact	85
VI/W/186 m	1.63	1.30	1.03	0.63	0.80	0.55	0.79	0.79	63	Compact bladed	79
VI/W/192 m	1.26	1.10	0.99	0.79	0.87	0.59	0.89	1.18	79	Compact	81
VI/W/198 m	1.35	1.10	0.86	0.64	0.81	0.51	0.79	0.16	64	Compact bladed	81
MEAN	1.26	1.06	0.87	0.70	0.84	0.53	0.83	0.36	70	Compact bladed	81

Note: L = Long, I = Intermediate, S = Short, F.R. = Flatness ratio, E.R. = Elongation ratio, MPSI = Maximum Projection Sphericity Index, OPI = Oblate Prolate Index.

1970). The pebble suites had maximum projection sphericity varying from 0.78–0.89 (mean = 0.83), which suggested product of a fluvial process. The coefficient of flatness ranged from 63%–79% (mean = 70%), which corroborate fluvial action as the dominant depositional process (Dobkin and Folk, 1970). The shape classes (Form) of these pebbles were Compact (C) and compact-bladed (CB) (Figure 13). These are diagnostic of fluvial environments (Dobkins and Folk, 1970) and it further supported dominant fluvial shaping for the grains. The upper limit for roundness index in fluvial environments has been placed at 45%. The roundness index for the Victoria Island pebbles range from 75–85%, with a mean of 81% suggesting a littoral environment. Most of other morphometric indices confirm that the Victoria Island pebbles were essentially deposited by fluvial processes.

5.4.3 Heavy mineral assemblages

The results from the three locations showed an abundance of staurolite. However, the mineral assemblage clearly suggests igneous to metamorphic rocks provenance. The ZTR index ranged from 44.3% – 58.6% (Victoria Island), 50.8% – 56.3% (Apapa), and 47.9% – 58.8% (Lekki) with an average overall ZTR index of 53.9%. Distribution bar chart of the ZTR index shows that all the analysed samples had greater than 40% ZTR index, which is suggestive of texturally immature sediments (Figures 9 and 10).

5.4.4 Petrography

The QFR ternary plot of mineral composition of the wells' samples (Figure 11) revealed that they are predominantly super mature quartz arenites with quartz ranging from 90 to 95% and feldspar 1 to 3% (Folk, 1974). They showed dominance of polycrystalline

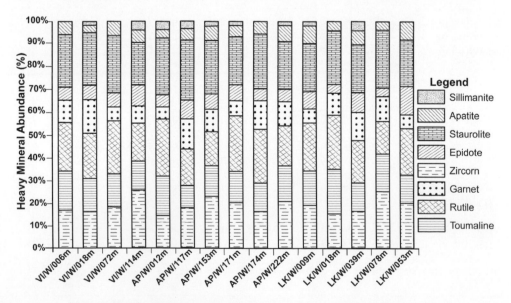

Figure 9. Relative abundance of heavy minerals in the three wells at different depths
(VI = Victoria Island, AP = Apapa, LK = Lekki).

quartz with most of the grains being angular to sub-angular; while the monocrystalline quartz are rounded to sub-rounded, indicating moderate to long distance of transportation from source (Tables 2 and 3; Figure 12).

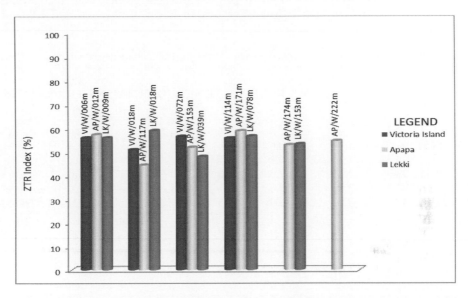

Figure 10. ZTR Index of the analysed samples from the studied wells at different depths (VI = Victoria Island, AP = Apapa, LK = Lekki).

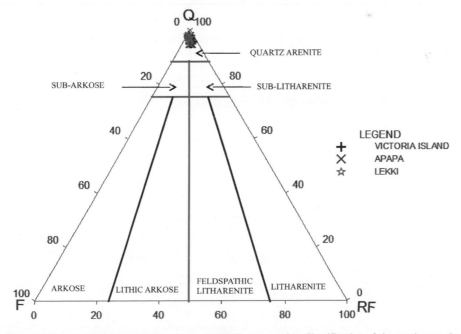

Figure 11. Quartz-Feldspar-Rock Fragments (QFRF) ternary plot: Classification of the sandy samples based on framework components.

5.4.5 Palynological findings

A total of 54 palynomorphs were recognized, 32 were pollen, six fern spores, one algae spore, and 15 fungal spores. Four main phyto-ecological groups were recognized, namely, mangrove swamp forest, freshwater swamp forest, freshwater, and lowland

Table 2. Average values of percentage composition of minerals in the analysed samples.

SAMPLE	QUARTZ %	FELDSPAR %	RUTILE %	GARNET %	TOURMALINE %	ROCK FRAGMENTS %
VI/W/006 m	92.5	1.5	1.0	0.5	1.5	3.0
VI/W/018 m	94.5	1.5	1.0	0.5	1.0	1.5
VI/W/072 m	92.5	1.5	1.5	0.5	2.5	1.5
VI/W/114 m	93.0	1.5	1.0	1.0	1.5	2.0
AP/W/012 m	92.5	2.0	1.0	0.5	2.0	2.0
AP/W/117 m	90.5	2.0	2.0	1.5	2.0	2.0
AP/W/153 m	90.5	2.5	1.5	1.5	1.5	2.5
AP/W/171 m	92.0	2.0	1.0	0.0	2.0	3.0
AP/W/174 m	93.5	1.5	1.0	1.5	0.5	2.0
AP/W/222 m	94.5	1.5	1.0	0.5	1.5	1.0
LK/W/009 m	90.0	1.5	1.5	1.0	2.0	4.0
LK/W/018 m	91.0	2.0	1.0	1.5	2.0	2.5
LK/W/039 m	92.5	1.5	0.5	1.0	2.0	2.5
LK/W/078 m	93.0	1.5	1.5	1.0	1.5	1.5
LK/W/153 m	92.0	2.5	0.5	1.0	1.5	2.5

Table 3. Quartz type, boundary relationship, and compositional maturity.

SAMPLE	AGE (ACCORDING TO*)	QUARTZ (GRAIN TYPE)	ANGULARITY /BOUNDRY	NO OF COUNT (AVERAGE)	COMPOSI-TIONAL MATURITY (AVERAGE)
VI/W/006 m	Recent	Polycrystalline	Angular to subangular	44.0	Super mature (= 92.5% qtz)
		Monocrystalline	Rounded to subrounded	21.0	
VI/W/018 m	Pleistocene to Oligocene	Polycrystalline	Angular to subangular	53.0	Supermature (= 94.5% qtz)
		Monocrystalline	Rounded to subrounded	24.0	
VI/W/072 m	Pleistocene to Oligocene	Polycrystalline	Angular to subangular	54.5	Supermature (= 92.5% qtz)
		Monocrystalline	Rounded to subrounded	22.0	
VI/W/114 m	Eocene	Polycrystalline	Angular to subangular	51.0	Supermature (= 93% qtz)
		Monocrystalline	Rounded to subrounded	20.5	

(Continued)

Table 3. *Continued.*

SAMPLE	AGE (ACCORDING TO*)	QUARTZ (GRAIN TYPE)	ANGULARITY /BOUNDRY	NO OF COUNT (AVERAGE)	COMPOSI-TIONAL MATURITY (AVERAGE)
AP/W/012 m	Pleistocene to Oligocene	Polycrystalline	Angular to subangular	39.5	Supermature (= 92.5% qtz)
		Monocrystalline	Rounded to subrounded	26.5	
AP/W/117 m	Eocene	Polycrystalline	Angular to subangular	41.5	Supermature (= 90.5% qtz)
		Monocrystalline	Rounded to subrounded	15.0	
AP/W/153 m	Palaeocene to Maastrichtian	Polycrystalline	Angular to subangular	31.0	Super mature (= 90.5% qtz)
		Monocrystalline	Rounded to subrounded	24.0	
AP/W/171 m	Palaeocene to Maastrichtian	Polycrystalline	Angular to subangular	41.0	Supermature (= 92% qtz)
		Monocrystalline	Rounded to subrounded	34.5	
AP/W/174 m	Palaeocene to Maastrichtian	Polycrystalline	Angular to subangular	37.0	Supermature (= 93.5% qtz)
		Monocrystalline	Rounded to subrounded	31.0	
AP/W/222 m	Lower Cretaceous	Polycrystalline	Angular to subangular	50.0	Supermature (= 94.5% qtz)
		Monocrystalline	Rounded to subrounded	26.0	
LK/W/009 m	Recent	Polycrystalline	Angular to subangular	38.5	Supermature (= 90% qtz)
		Monocrystalline	Rounded to subrounded	27.0	
LK/W/018 m	Pleistocene to Oligocene	Polycrystalline	Angular to subangular	52.0	Supermature (= 91% qtz)
		Monocrystalline	Rounded to subrounded	30.5	
LK/W/039 m	Pleistocene to Oligocene	Polycrystalline	Angular to subangular	47.0	Supermature (= 92.5% qtz)
		Monocrystalline	Rounded to subrounded	29.5	
LK/W/078 m	Pleistocene to Oligocene	Polycrystalline	Angular to subangular	40.5	Supermature (= 93% qtz)
		Monocrystalline	Rounded to subrounded	16.0	
LK/W/153 m	Palaeocene to Maastrichtian	Polycrystalline	Angular to subangular	50.5	Super mature (= 92% qtz)
		Monocrystalline	Rounded to subrounded	23.5	

(*Jones and Hockey 1964; Reyment 1965; Ogbe, 1972; Billman 1980; Omatsola and Adegoke1981; Ajakaiye and Bally 2002; Brownfield and Charpentier 2006).

rainforest. Others with no clearly distinctive groups were classified under their respective plant types. These include open vegetation, ferns, and fungi (Tables 4 and 5).

In Group A, the mangrove swamp forest species were abundant with most dominant of *Rhizophora* spp. and fungi freshwater swamp forest, lowland rainforest species (LRF), ferns and open vegetation dominated by *Monoporites annulatus*. The abundance of *Rhizophora* spp. indicates predominance of mangrove swamp forest with fairly high sea level (Rull, 1998; Poumot, 1989). Occurrence of freshwater swamp forest species such as *Leea guineensis* and *Mitragyna ciliata* suggest influx of freshwater; while the occurrence of LRF pollen, *Alchornea* sp., *Bombacacidites* sp., those genera of open vegetation and spores of forest-dwelling ferns show some input from continental landscapes corroborated by the occurrence of *Alchornea* sp., a pioneer species.

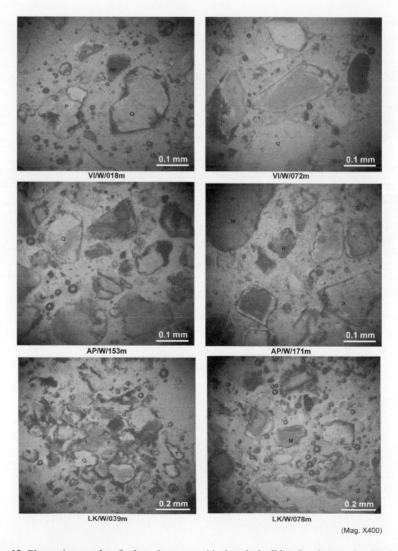

(Mag. X400)

Figure 12. Photomicrographs of selected petrographical analysis slides. Q = Quartz, P = Plagioclase, M = Micas, T = Tourmaline, R = Rutile, G = Garnet, RF = Rock Fragments (Mag. ×400).

Group B mangrove swamp forest species especially *Rhizophora* spp. was reduced (7.7%). *Rhizophora* spp. constitutes 40–90% of the pollen sum in the mangrove swamp forest of Nigeria (Sowunmi, 1981; Orijemie, 2013). Occurrence of less than 40% of *Rhizophora* spp. has been reported to indicate a much reduced mangrove swamp forest and lowered sea level, which is corroborated by the good representation of freshwater swamp forest species (FWSF), LRF, ferns, fungi, and *Botryococcus* fresh water algae (Sowunmi, 1987). These assemblages indicate environment of deposition with little influence from brackish waters and the occurrence of abundant FWSF species (*Arecipites* sp., *Mitragyna ciliata, Symphonia globulifera,* and *Uapaca* sp.), and *Botryococcus* strongly suggest freshwater influx. The fair representation of LRF (*Diospyros* sp. and *Tricolporopollenites* sp.), ferns, and fungi suggest deposition near the shore.

In Group C, the mangrove swamp forest was dominant and represented by *Rhizophora* spp., *Avicennia Africana, Laguncularia racemosa,* the salt water fern, *Acrostichum aureum* (Rull, 1998), FWSF (*Mitragyna ciliata, Uapaca* sp., *Proxapertites* sp., *Retimonocolpites,* that is, palms, *Typhaaustralis*), and LRF (*Bombax buonopozense, Carapa procera, Celtis* sp., *Diospyros, Pentaclethra macrophylla* and *Retitricolporites irregularis*). These assemblages suggest environment dominated by mangroves, rainforests, and palm and freshwater swamp forest, deposited mainly in a coastal environment with occasional influx from rivers located in the hinterland (Tables 4 and 5, Figure 13).

Table 4. Palynomorphs representation in the well samples at different depths.

Palynomorphs/ Samples	LK/W/ 3 m	LK/W/ 51 m	LK/W/ 21 m	LK/W/ 39 m	VI/W/ 42 m	VI/W/ 54 m	AP/W/ 117 m	AP/W/ 153 m	AP/W/ 174 m
Acanthaceae	0	0	0	0	0	0	0	0	1
Acrostichum aureum	0	0	0	2	7	0	0	0	1
Alchornea sp.	0	0	0	2	0	0	0	0	0
Arecidites sp.	1	0	0	0	3	0	0	2	0
Avicennia africana	0	0	0	0	0	0	0	2	0
Bombacacidites sp.	0	0	0	1	0	0	0	0	0
Bombax buonopozense	0	0	0	0	0	4	0	0	0
Botryococcus sp.	0	2	0	0	3	0	0	0	0
Carapa procera	0	0	0	0	0	0	0	1	0
Celtis sp.	0	0	0	0	0	0	0	1	0
Cercophora sp.	0	3	0	0	0	0	0	0	0
Cissus quadrangularis	0	0	0	0	0	3	0	0	0
Clasterosporium sp.	0	2	0	0	0	0	0	0	0
Combretaceae/ Melastomataceae	0	0	0	2	2	0	0	2	0
Daniellia sp.	0	0	0	1	0	0	0	0	0
Dicellaesporites africanus	0	0	1	0	0	0	0	0	0
Dictosporites camerounensis	0	0	1	0	0	0	0	0	0
Dictyosporites disctyosus	0	0	1	0	3	0	0	0	0
Dictyosporites heterospora	0	0	0	0	6	0	0	0	0

(Continued)

Table 4. *Continued.*

Palynomorphs/ Samples	LK/W/ 3 m	LK/W/ 51 m	LK/W/ 21 m	LK/W/ 39 m	VI/W/ 42 m	VI/W/ 54 m	AP/W/ 117 m	AP/W/ 153 m	AP/W/ 174 m
Dictysporites africanus	0	0	0	0	6	0	0	0	0
Didymoporispororites mucronatus	0	0	0	0	0	0	0	0	1
Diospyros sp.	0	0	0	0	4	0	0	2	0
Diporicellaesporites anisospora	0	0	0	0	0	1	0	0	0
Gnetaceaepollenites sp.	0	0	0	0	0	0	0	1	0
Euphorbiaceae	0	0	0	0	0	0	0	1	0
Fusiformisporites pseudocrabbii	0	3	0	0	0	0	0	0	0
Hypoxylonites xylarioides	0	0	0	0	2	0	0	0	0
Laevigatosporites sp.	0	0	0	0	0	0	0	3	2
Lasiodiplodia theobromae	0	2	0	0	0	0	0	0	0
Leea guineensis	0	0	0	1	0	0	0	0	0
Lonchorcarpus sp.	0	0	0	0	0	0	0	1	0
Mitragyna ciliata	0	9	0	3	3	0	1	0	0
Monolete spore	1	0	0	1	0	0	0	0	0
Monolete spore (Large)	0	7	0	0	0	0	0	0	0
Monoporites annulatus	3	3	0	0	0	0	0	1	0
Multicellaesporites crassispora	0	0	0	0	0	4	0	0	0
Multicellaesporites doualae	0	0	0	0	0	3	0	0	0
Multicellaesporites sp.	0	6	0	0	0	0	0	0	0
Myrtaceae	0	0	0	0	0	0	0	2	0
Pachydermites diederixii	0	0	0	0	3	0	0	0	0
Pentachletra macrophylla	0	0	0	0	0	0	0	1	0
Phyllanthus sp.	0	0	0	0	0	4	0	0	1
Polypodiaceoisporites sp.	0	2	0	0	3	0	0	0	0
Proxapertites sp.	0	0	0	0	0	7	0	0	0
Pycnanthus angolensis	0	0	0	0	0	2	0	0	0
Retimonocolpites sp.	0	0	0	0	0	0	0	2	0
Retitricolporites irregularis	0	0	0	0	0	0	0	1	0
Rhizophora sp.	8	1098	41	91	12	843	3	64	29
Sapotaceae	0	0	0	0	0	0	0	1	0
Tabernaemontana crassus	0	0	0	0	0	0	0	1	0
Tricolporopollenites sp.	0	2	0	0	3	0	0	0	0
Typha cf. australis	0	0	0	0	0	0	0	2	0
Uapaca sp.	0	0	0	0	11	3	0	0	1
Unidentified	0	0	0	0	0	6	0	0	0
Unidentified	0	0	0	0	5	0	0	0	0
Unidentified 2	0	0	0	0	0	2	0	0	0
Verrucatosporites usmensis	0	0	0	0	2	0	0	0	0
Total	13	1139	44	104	78	882	4	91	36

Table 5. Phyto-Ecological Groups recognized in the well samples.

Mangrove Swamp Forest	Freshwater Swamp Forest	Freshwater
Acrostichum aureum (Fern)	*Arecipites* sp. (Arecaceae [Palmae])	*Botryococcus* sp.
Avicennia africana	*Cissus quadrangularis*	*Typha australis*
Laguncularia racemosa	*Leea guineensis*	
Rhizophora sp.	*Mitragyna ciliata*	
	Pachydermites diederixi (*Symphonia globulifera*)	
	Proxapertites sp. (Arecaceae [Palmae])	
	Retimionocolpites sp. (Arecaceae [Palmae])	
	Uapaca sp.	

Lowland Rainforest	Fungi	Open Vegetation
Alchornea sp. (*Psilatricolporites operculatus*)	*Cercophora* sp.	*Daniellia oliveri* (*Psilatricolporites* sp.)
Amanoa sp. (*Retitricolporites irregularis*)	*Clasterosporium* sp.	*Gnetaceaepollenites* sp.
Bombacacidites sp.	*Dicellaesporites africanus*	*Phyllanthus* sp. (*Retitricolporites* sp.)
Bombax buonopozense (*Bombacacidites* sp.)	*Dictosporites camerounensis*	Poaceae (*Monoporites annulatus*)
Carapa procera (*Tricolporites* sp.)	*Dictyosporites disctyosus*	
Celtis sp. (*Tricolporites* sp.)	*Dictyosporites heterospora*	
Diospyros sp. (*Psilatricolporites benueensis*)	*Dictyosporites africanus*	
Pentaclethra macrophylla (*Brevicolporites guinetii*)	*Didymoporispororites mucronatus*	
Pycnanthus angolensis	*Diporicellaesporites anisospora*	
Tabernaemontana crassa (*Psilatricolporites crassus*)	*Fusiformisporites pseudocrabbii*	
Tricolporopollenites sp.	*Hypoxylonites xylarioides*	
	Lasiodiplodia theobromae	
	Multicellaesporites crassispora	
	Multicellaesporites doualae	
	Multicellaesporites sp.	

Ferns
Laevigatosporites sp.
Monolete spores
Polypodiaceoisporites sp.
Verrucatosporites usmensis

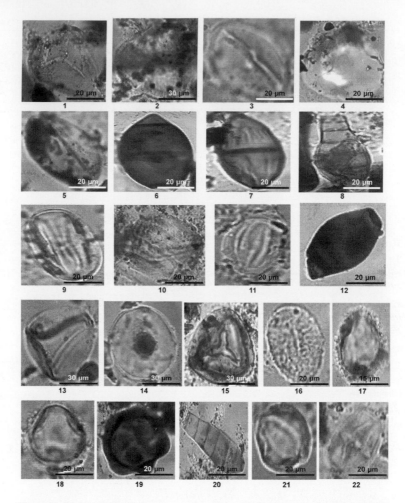

Figure 13. Photographs of the recovered and identified palynomorphs (Mag. ×1000).

(1) *Acrostichum aureum*—AP/W/174 m (Dorhofer)

(2) *Acanthaceous* sp.—AP/W/174 m

(3) *Avicennia* sp.—AP/W/153 m

(4) *Bombacacidites* sp. (*Bombax buonopozense*) —LK/W/039 m

(5) *Cissus* sp.—VI/W/054 m

(6) *Didymoporispororites mucronatus*—AP/W/174 m (Salard-Cheboldaeff and Locquin, 1980)

(7) *F. psuedoscrabbii*—LK/W/051 m

(8) *Diporicellaesporites anispora* (Salard-Cheboldaeff and Locquin, 1980)

(9) *Diospyros* sp.—AP/W/153 m

(10) *Ephedra*—AP/W/153 m

(11) *Lonchorcarpus* sp.—AP/W/153 m

(12) *M. carssifora*—VI/W/054 m

(13) *Monoporites annulatus*—LK/W/051 m (Van der Hammen, 1954)

(14) *Pentachlethra macrophylla*—AP/W/153 m (Salard-Cheboldaeff, 1980)

(15) *Pteris* sp. 2 – VI/W/042 m

(16) *Pycnanthus angolense*—VI/W/054 m

(17) *Retitricolporites irregularis*—AP/W/153 m (Van der Hammen and Wymstra, 1964)

(18) *Uapaca* sp.—AP/W/153 m

(19) Unidentified sp.—VI/W/054 m

(20) Fungi spore—LK/W/039 m

(21) *Rhizophora* sp.—AP/W/153 m

(22) *Rhizophora* sp.—VI/W/054 m

5.4.6 Stable isotopes and possible palaeoclimatic implications

The results of the oxygen and carbon isotopes are presented in Tables 6 and 7. Specimen 1, a gastropoda belonging to the genus *Neptunea*, a species of *Neptunea tabulate,* and Specimen 2, a bivalves species of *Venericardia planicosta,* were recovered from diverse depths (Lekki well). They showed oxygen isotopic compositions of −2.74‰ and −2.94‰, while the carbon isotopic compositions are −6.07‰ and −7.78‰. The enrichment of the gastropods relative to bivalves in both oxygen and carbon isotopic composition suggested species-specific fractionation effects (vital effects) since they were formed within the same ambient water. Gastropods (*Neptunea tabulata*) are 0.2‰ enriched in $\delta^{18}O$ and 1.71‰ enriched in $\delta^{13}C$ than bivalves (*Venericardia planicosta*). Within the same locality and stratigraphic level, different molluscs species clearly exhibited different isotope signatures. These were caused by habitat segregation and local variations in the lake environment (Harzhauser *et al.*, 2011).

The $\delta^{18}O$ isotopic composition of the *Venericardia planicosta* (Specimens 2 to 6) ranged from −4.93‰ to −2.94‰ VPDB (with mean value = 4.22‰) and the $\delta^{13}C$ isotopic composition range from −10.35‰ to −3.71‰ VPDB (mean value = −7.43‰).

Table 6. Lekki well.

Specimen Numbers	Specimen Depth Range	Type of Specimen	Age (According to *)	$\delta^{13}C$ Carbonate ‰ (VPDB)	$\delta^{18}O$ Carbonate ‰ (VPDB)	EFPT °C
1	LK/W/ 3–30 m(G)	*Neptunea tabulata*	Pleistocene to Oligocene	−6.07	−2.74	31.32
2	LK/W/ 3–30 m(B)	*Venericardia planicosta*	Pleistocene to Oligocene	−7.78	−2.94	32.19
3	LK/W/ 33–51 m	*Venericardia planicosta*	Pleistocene to Oligocene	−10.35	−4.93	40.82
4	LK/W/ 108–153 m	*Venericardia planicosta*	Palaeocene to Maastrichtian	−6.47	−4.60	39.39
5	LK/W/ 156–198 m	*Venericardia planicosta*	Palaeocene to Maastrichtian	−8.82	−4.92	40.78
6	LK/W/ 207–246 m	*Venericardia planicosta*	Palaeocene to Maastrichtian	−3.71	−3.72	35.57

*Jones and Hockey, 1964; Reyment, 1965; Ogbe, 1972; Billman, 1980; Omatsola and Adegoke, 1981; Ajakaiye and Bally, 2002; Brownfield and Charpentier, 2006.

Table 7. Apapa well.

Specimen Number	Specimen Depth Range	Type of Specimen	Age (According to *)	$\delta^{13}C$ Carbonate ‰ (VPDB)	$\delta^{18}O$ Carbonate ‰ (VPDB)	EFPT °C
7	AP/W/ 15–27 m	*Venericardia planicosta*	Pleistocene to Oligocene	−3.27	−2.48	30.19
8	AP/W/ 30–36 m	*Venericardia planicosta*	Pleistocene to Oligocene	−1.96	−2.19	28.93

EFPT = Estimated Formational Palaeotemperature.

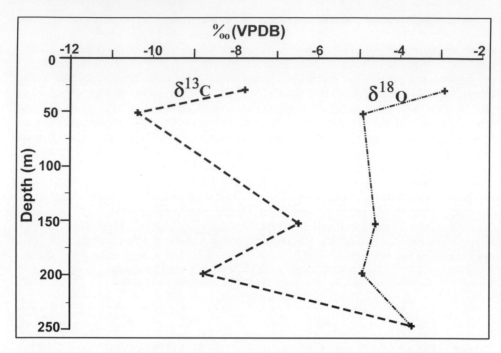

Figure 14. Plot of depth against oxygen and carbon isotopic composition of *Venericardia planicosta* from Lekki well.

The range of $\delta^{13}C$ values are characteristic of freshwater species (Harzhauser *et al.*, 2011). The variation in oxygen isotopic composition is a reflection of the fluctuations in temperature and seasonal changes. The moderately negative values of oxygen isotopic composition of the *Venericardia planicosta* suggests enough freshwater contribution that keeps the oxygen values close to –5‰. The strongly negative carbon values suggest that the *Venericardia planicosta* utilized a mixed carbon pool of dissolved inorganic carbon (high $\delta^{13}C$ of between –3 to +3‰) directly from the ambient water (McConnaughey and Gillikin, 2008) and dietary carbon (metabolic carbon) during their shell formation (Dettman, 1999; Geist *et al.*, 2005; Gajurel *et al.*, 2006). The oxygen and carbon isotopic composition of the *Venericardia planicosta* against depth shows a positive correlation between oxygen and carbon isotopes of the carbonates (Figure 14). Lower $\delta^{13}C$ concentrations are associated with lower $\delta^{18}O$ and conversely, higher $\delta^{13}C$ values co-occur with higher $\delta^{18}O$ values. Some co-variation between $\delta^{18}O$ and $\delta^{13}C$ suggest that seasonal climate change might influence carbon source to some degree (Leng and Lewis, 2014; Geist *et al.*, 2005; Goodwin *et al.*, 2001). The cross-plot of oxygen and carbon isotopic composition of the *Venericardia planicosta* showed the trend of changes in the isotopic composition of shell carbonate during formation as illustrated by the arrow (Figure 15). However, significant relationships do exist between the shell isotopic composition and shell sizes relative to varying climatic changes. The Apapa well oxygen and carbon isotopic composition showed higher negative values. The $\delta^{18}O$ values are –2.48‰ and –2.19‰, while $\delta^{13}C$ values are –3.27‰ and –1.96‰. The higher values in both stable isotopes suggest a combination of evaporation effects and the influx of already isotopically heavy freshwater from the catchment area (Harzhauser *et al.*, 2011; Geist *et al.*, 2005; Goodwin *et al.*, 2001).

Evaporation and precipitation play a significant role in isotopic enrichment and depletion of ambient water. In freshwater molluscs, $\delta^{18}O$ often preserve changes in the

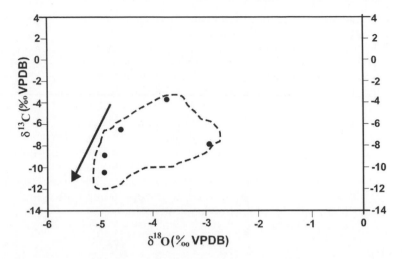

Figure 15. Cross-plots of oxygen and carbon isotopic composition of *Venericardia planicosta* from Lekki well. Trend shown by arrow illustrate changes in the isotope composition of the shell carbonates during formation.

lake water. Low latitude (closed) lakes tend to have lake water oxygen isotope variations that are dominated by dry season evaporation (high $\delta^{18}O$) and wet season recharge (low $\delta^{18}O$); while at high latitudes in open lake systems, lake water $\delta^{18}O$ is often controlled by the amount, source, and temperature of precipitation (Leng *et al.*, 2001; Bar-Yosef *et al.*, 2012). Plots of $\delta^{18}O$ against depth illustrate changes from high to low and high $\delta^{18}O$ values in response to variable climate. The high values indicate dry season evaporation; while the low values are the wet seasonal recharge. Hence, the shell carbonates were formed within variable climatic changes. The palaeotemperature was estimated using Grossman and Ku (1986) equation. It ranged from 28.93°C–40.82°C. The $\delta^{18}O$ water of 0‰ ($\delta^{18}O$ of freshwater) was used in the estimation of the palaeotemperature (Cronin, 1999). Due to this assumed variable, these estimations are only approximations and have to be treated with caution.

5.5 CONCLUSIONS

The sandstone recovered from the three wells comprises of siltstone, sandstone, claystone, and pebbly conglomerates. The granulometric studies show that the sandstones are medium to very coarse grained, moderately well sorted to very poorly sorted. They show strongly positive skewness (near symmetrical) and a wide range of kurtosis from very platykurtic to extremely leptokurtic. The ratio of standard deviation (sorting) against skewness and sorting against mean sizes indicate that the sandstones were river deposit under medium to high energy. The heavy mineral assemblage and average ZTR index in the area suggest that the sediments are texturally immature from igneous and metamorphic sources. The studies have shown that the sandstones are supermature quartz arenites and are compositionally mature due to high percentage of quartz grains with little or no feldspar. They show dominance of polycrystalline quartz with most of the grains being angular to sub-angular, while the monocrystalline quartz are rounded to sub-rounded rounded indicating moderate to long distance of transportation. The palynological study suggests abundance of mangrove swamp forest species, with the dominance of

Rhizophora sp. and the occurrence of abundant freshwater species, which suggests high influx of freshwater from rivers. Morphometry of quartz pebbles from Victoria Island well have shown maximum projection sphericity indicating product of a fluvial process. Compact and compact-bladed forms of the pebbles are diagnostic of fluvial environment, which is corroborated by the coefficients of flatness of the pebbles. The roundness of the pebbles strongly suggests deposition in littoral environment. Isotopic studies have shown oxygen and carbon isotopic composition to be characteristics of variable climatic changes. The $\delta^{13}C$ values are typical of freshwater environment and the variation in oxygen isotope composition is a reflection of the fluctuations in temperature, which also determines the seasonal changes. The strongly negative carbon isotope composition of shells from Lekki well revealed that the genus *Neptunea* utilized both the dissolved inorganic carbon from the ambient water and metabolic carbon in their shell formation. The moderately negative values of both isotopes in shells from Apapa well suggest evaporation effects and influx of isotopically heavy freshwater from the hinterland catchment area.

REFERENCES

Adegoke, O.S., 1969, Eocene stratigraphy of southern Nigeria. Colloque sur L'Eocene 111, *Bureau Recherche Géologiques et Minières*, **69**, pp. 23–48.

Adegoke, O.S., 1977, Stratigraphy and palaeontology of the Ewekoro Formation (Paleocene) of South Western Nigeria. *Bulletins of American Paleontology*, **71**, pp. 295.

Agagu, O.A., 1985, *A Geological guide to Bituminous Sediments in South Western Nigeria*. unpubl. Report, Dept. of Geology, University of Ibadan, Ibadan.

Ajakaiye, D.E. and Bally, A.W., 2002, Some Structural Styles on Reflection Profiles from Offshore Niger Delta. *American Association of Petroleum Geoscientist, Education Course Note Series*, **41**, pp 3–4.

Akinmosin, A. and Osinowo, O.O., 2008, Geochemical and Mineralogical Composition of Ishara Sandstone Deposit, Southwestern Nigeria. *Continental Journal of Earth Sciences*, **3**, pp. 33–39.

Ako, B.D., Adegoke, O.S. and Petters, S.W., 1981, Stratigraphy of the Oshosun Formation in South Western Nigeria. *Journal of Mining and Geology*, **17**, pp. 97–106.

Bar-Yosef Mayer, D.E., Leng, M.J., Aldridge, D.C., Arrowsmith, C., Gumus, B.A. and Sloane, H.J., 2012, Modern and early-middle Holocene shells of the freshwater molluscs from Catalhoyuk in the Konya Basin, Turkey: Preliminary palaeoclimatic implications from molluscan isotope data. *Journal of Archaeological Science*, **39**, pp. 76–83.

Bassey, C.E., 1996, Grain size analysis and heavy mineral composition of Beach sand in Victoria, *African Journal of Science and Technology,* Series B, **8**(1), pp. 11–14.

Billman, H.G., 1976, Offshore stratigraphy and paleontology of Dahomey embayment, West Africa. *Proc. 7th African Micropaleontology Coll., Ile-Ife*, pp. 27–42.

Boboye, O.A. and Nwosu, O.R. 2013, Petrography and geochemical indices of the Lagos lagoon coastal sediments, Dahomey Basin (southwestern Nigeria): Sea level change implications. *Quaternary International Journal*, **338**, pp. 14–27.

Brownfield M.E. and Charpentier, R.R., 2006, Geology and total petroleum systems of the West-Central Coastal Province (7203), West Africa, *U.S. Geological Survey Bulletin*, **2207-C**, pp. 52.

Cronin, T.M., 1999, *Principles of Palaeoclimatology*, (New York: Columbia University Press), pp. 75–86.

Dettman, D.L., 1999, Controls on stable isotope composition of seasonal growth bands in aragonitic freshwater bivalves (unionidea). *Geochimica et Cosmochimica Acta, 63*, pp. 1049–1057.

Dobkins, J.E. and Folk, R.L., 1970, Shape development on Tahiti-Nui. *Journal of Sedimentary Petrology*, **40**, pp.1167–1203.

Els, B.G., 1988, Pebble Morphology of an ancient conglomerate. The Middelvlei Gold Placer, Witwatersrand, South Africa. *Journal of Sedimentary Petrology*, **58**(5), pp. 894–902.

Elueze, A.A. and Nton, M.E., 2004, Composition Characteristics and Industrial Assessment of Sedimentary Clay bodies impact of Eastern Dahomey Basin, South Western Nigeria. *Journal of Mining and Geology,* **4**(12), pp. 175–184.

Elvsborg, A. and Dalode, J., 1985, Benin hydrocarbon potential looks promising. *Oil Gas Journal*, **83**, pp. 126–131.

Evamy, B.D., Herembourne, J., Kameling, P., Knap, W.A., Molly, F.A. and Rowlands, P.H., 1978, Hydrocarbon habitat of Tertiary Niger Delta, *American Association of Petroleum Geologists Bulletin*, **62**, pp. 1–39.

Folk, R.L., 1974, *Petrology of sedimentary rocks*, (Austin: Hemphil Publishing Company), P182.

Friedman, G.M., 1967. Distinction between Dune, Beach and river sands from textural characteristics. *Journal of Sedimentary Petrology,* **31**, pp. 514–529.

Gajurel, A.P., France-Lanord, C., Huyghe, P., Guilmette, C. and Gurung, D., 2006, Carbon and oxygen isotope compositions of modern freshwater mollusc shells and river waters from the Himalayas and Ganga plain. *Chemical Geology,* **233**, pp. 156–183.

Geist, J., Auerswald, K. and Boom, A., 2005, Stable carbon isotopes in freshwater mussel shells: environmental record or marker for metabolic activity. *Geochimica et Cosmochimica Acta,* **69**, pp. 3545–3554.

Germeraad, J.H., Hopping, C.A. and Muller, J. (1968), Palynology of Tertiary sediments from tropical areas. *Review of Palaeobotany and Palynology*, **6**(3–4), pp.189 348.

Goodwin, D.H., Flessa, K.W., Schone, B.R. and Dettman, D.L., 2001, Cross-calibration of daily growth increments, stable isotope variation, and temperature in the gulf of California bivalve mollusc *Chione cortezi*: Implications for palaeoenvironmental analysis. *Palaios,* **16**, pp. 387–398.

Grossman, E.L. and Ku, T., 1986, Oxygen and carbon isotope fractionation in biogenic arogonite: Temperature effects. *Chemical Geology*, **59**, pp. 59–74.

Harzhauser, M., Mandic, O., Latal, C. and Kern, A., 2011, Stable isotope composition of the Miocene Dinaride Lake System deduced from its endemic mollusc fauna. *Hydrobiologia*, **682**(1), pp. 27–46. doi: 10.1007/s10750-011-0618-3.

Jones, H.A. and Hockey, R.D., 1964, The geology of part of southwestern Nigeria. *Geological Survey Bulletin*, **31**, pp. 1–101.

Kingston, D.R., Dishroon, C.P. and Williams, P.A., 1983a, Global basin classification. *American Association of Petroleum Geologists, Bulletin,* **67**, pp. 2175–2193.

Kingston, D.R., Dishroon, C.P. and Williams, P.A., 1983b, Hydrocarbon plays and global basin classification. *American Association of Petroleum Geologists, Bulletin*, **67**, pp. 2194–2198.

Klemme, H.D., 1975, Geothermal Gradients, Heatflow and Hydrocarbon Recovery. In *Petroleum and Global Tectonics*, edited by Fischer, A.G. and Judson, S., (Princeton, New Jersey: Princeton University Press), pp. 251–306.

Krumbein, W.C., 1941, Measurements and geological significance of shape and roundness of sedimentary particles. *Journal of Sedimentary Petrology*, **11**, pp. 64–72.

Leng, M.J. and Lewis, J.P. 2014, Oxygen isotopes in molluscan shell: Applications in environmental archaeology. *Environmental Archaeology*, **21**(3), pp. 295–306.

Leng, M.J., Barker, P., Greenwood, P., Roberts, N. and Reed, J.M. 2001, Oxygen isotope analysis of diatom silica and authigenic calcite from Lake Pinarbasi, Turkey. *Journal of Paleolimnology*, **25**, 343–9.

Lutig, G.O., 1962, The shape of pebbles in the continental, fluviatile and marine facies. *International Association Scientific Hydrology Publication*, **59**, pp. 234–258.

McConnaughey, T.A. and Gillikin, D.P., 2008, Carbon isotopes in mollusc shell carbonates. *Geo-Marine Letters*, **28**, pp. 287–289.

Molodkov, A.N. and Bolikhovskaya, N.S., 2002, Eustatic sea-level and climate changes over the last 600 ka as derived from mollusc-based ESR-chronostratigraphy and pollen evidence in Northern Eurasia. *Sedimentary Geology*, **150**, pp. 185–201.

Ogbe, F.A.G., 1972, Stratigraphy of strata exposed in the Ewekoro quarry, southwestern Nigeria. In *African Geology*, edited by Dessauvagie, T.F.J. and Whiteman A.J. (Ibadan: University of Ibadan Press), pp. 305–322.

Okosun, E.A., 1987, Ostracod biostratigraphy of the Eastern Dahomey Basin, Niger Delta and Benue Trough of Nigeria. *Bulletin Geological Survey of Nigeria*, **41**, pp. 151.

Omatsola, M.E. and Adegoke, O.S., 1981, Tectonic evolution and Cretaceous stratigraphy of the Dahomey Basin, Nigeria. *Journal of Mining and Geology*, **18**(1), pp. 30–137.

Orijemie, E.A., 2013, A palynological and archaeological investigation of the environment and human occupation of the Rainforest of southwestern Nigeria during the Late Holocene. Unpublished PhD Thesis, Department of Archaeology and Anthropology, University of Ibadan, pp. 313.

Poumot, C., 1989, Palynological evidence for eustatic events in the Tropical Neogene. *Bull. Cent. Res.- Exp. Prod. Elf-Aquitaine*, **13**(2), pp. 437–453.

Rull, V., 1998, Middle Eocene mangroves and vegetation changes in the Maracaibo Basin, Venezuela. *Palaios,* **13**, pp. 287–296.

Salzmann, U., Hoelzmann, P. and I. Morczinek 2002, Late Quaternary climate and vegetation of the Sudanian zone of NE-Nigeria deduced from pollen, diatoms and sedimentary geochemistry. *Quaternary Research,* **58**, pp. 73–83.

Sames, C.W., 1966, Morphometric data of some recent pebble association and their application to ancient deposits. *Journal of Sedimentary Petrology*, **36**, pp. 126–142.

Schone, B.R., Freyre Castro, A.D., Fiebig, J., Houk, S.D., Oschamann, W. and Kroncke, I., 2004, Sea surface water temperature over the period 1884–1983 reconstructed from oxygen isotope ratios of a bivalve mollusc shell (Arctica islandica, Southern North Sea). *Palaeogeography, Palaeoclimatology, Palaeoecology,* **212**, pp. 215–232.

Sneed, E.D. and Folk, R.L., 1958, Pebbles in the lower Colorado River, Texas; A study in particle morphogenesis. *Journal of Geology*, **66**, pp. 114–150.

Sowunmi, M.A., 1981, Late Quaternary environmental changes in Nigeria, *Pollen et Spores*, **23**(1), pp. 125–148.

Sowunmi, M.A., 1987, Palynological studies in the Niger Delta. In *The early history of the Niger Delta*, edited by Alagoa, E.J., Anozie, F.N. and Nzewunwa, N., (Hamburg: Helmut Buske Verlag), pp. 29–59.

Sowunmi, M.A., 2004, Aspects of Nigerian coastal vegetation in the Holocene: some recent insights. In *Past climate through Europe and Africa*, edited by Battarbee, R.W., Gasse, F. and Stickley, C.E., (Dordrecht: Springer), pp. 199–218.

Stratten, T., 1974, Notes on the Application of Shape Parameters to Differentiate between Beach and River Deposits in Southern Africa. *South African Journal of Geology*, **77**(1), pp. 59–64.

CHAPTER 6

Palynofacies, sedimentology and palaeoenvironment evidenced by studies on IDA-6 well, Niger Delta, Nigeria

Jacinta N. Chukwuma-Orji, Edward A. Okosun,
Isah A. Goro & Salome H. Waziri
*Department of Geology, Federal University of Technology,
Minna, Nigeria*

ABSTRACT: Palynofacies and sedimentological analyses of the strata penetrated by Ida-6 well were undertaken with the aim of determining biozones and the palaeoenvironmental conditions. Fifty ditch cutting samples within the intervals of 2679–4051 m were analysed. The analysis yielded low to abundant recovery of pollen and spores, small to large sizes of palynomaceral 1 (irregularly shaped, orange-brown to dark-brown coloured and opaque plant debris), palynomaceral 2 (irregularly shaped, brown-orange coloured platy plant materials), and few occurrences of palynomaceral 3 (pale to brown coloured, cuticular, and translucent plant materials). The lithology showed alternation of shale and sandstone units with few intercalations of siltstone units. The sandstone units consist of fine to medium grains, occasionally coarse to granule sized. The sand grains are mostly sub-angular to sub-rounded, occasionally rounded, and generally poorly to moderately sorted. The accessory minerals are dominated by ferruginous material, shell fragments and carbonaceous detritus with spotty occurrences of mica flakes. The lithologic, textural, and wire line log data indicated that the entire studied intervals in the well belong to the Agbada Formation. The studied intervals were deposited during middle Miocene to late Miocene based on the recovered age diagnostic marker species such as *Multiareolites formosus, Verrutricolporites rotundiporus, Crassoretitriletes vanraadshoveni* and *Racemonocolpites hians*. Three interval range zones were established using the international stratigraphic guide for the establishment of biozones. The three established palynostratigraphic zones are *Multiareolites formosus–Lavigatosporites* sp., *Racemonocolpites hians–Crassoretitriletes vanraadshoveni* and *Psiltricolporites crassus—Acrostichum aureum* Zones. Coastal-deltaic environments of deposition have been inferred for the studied interval of the well on the bases of the palynofacies association and sedimentological characteristics.

6.1 INTRODUCTION

Palynofacies has been described to mean the total organic matter that is recovered from a rock or unconsolidated sediment by the standard palynological processing technique of digesting samples in HCl and/or HF (Batten and Stead, 2005). In palynofacies analysis, it is not only the palynomorphs (pollen, spores, dinoflagellate cysts, fungal remains, and foraminiferal linings) in the palynological slides that are investigated, but the entire organic content of the slides. The organic content includes structured organic matter (phytoclasts and zooclasts) and unstructured organic matter which comprises of amorphous organic matter, gelified matter, solid bitumen, resin, and amber (Batten and Stead, 2005). The nature of the organic contents in the sedimentary rocks reflects the original conditions in the source area as well the depositional environments. The integration of palynofacies and sedimentological analyses significantly contribute to environmental reconstruction and basin evaluation (Oyede, 1992). The understanding of the biochronology and palaeoenvironment are essential in basin evaluation and successful exploration of both organic and inorganic mineral resources. Palynology, sedimentology, and palaeoenvironmental conditions of sedimentary rock deposition within the Niger Delta basin have been published by Ige *et al.* (2011), Ajaegwu *et al.* (2012), Ojo and Adebayo (2012), and Olajide *et al.* (2012). The most comprehensive contribution to the knowledge on the palynology of the Niger Delta was made by Germeraad *et al.* (1968). The study was based on the palynomorph assemblages of the Tertiary sediments of three tropical areas: parts of South America, Asia and Africa (Nigeria). They established nine pantropical zones using quantitative base and top occurrence (numeric method) of diagnostic species such as *Echitricolporite spinosus, Crassoretitriletes vanradshoveni, Magnastrites howardi, Verrucatosporites usmensis, Monoporites annulatus,* and *Proxapertites operculatus.* Evamy *et al.* (1978) established twenty-nine informal palynological zones of the Niger delta using alphanumeric coding method, which seems to form the background information for in-house zonal scheme of Shell Petroleum Development Company. Palynological studies of sediments from North Chioma-3 Well, Niger Delta and its palaeoenvironmental interpretations were carried out by Ige *et al.* (2011). They recognized four pollen zones (I-IV), using pollen diagram method. They further established both wet and warm climate using percentage occurrence of mangrove forest taxa. They interpreted the palaeoenvironment as mangrove swamp environment because of high occurrence of *Rhizophora* sp. palynology of Bog-1 well, south-eastern Niger Delta was studied by Olajide *et al.* (2012). They noted that dominance occurrence of the mangrove species, *Zonocostites ramonae* (*Rhizophora*) and *Foveotricolporites crassiexinus* (*Avicennia*), suggests a tidal swamp shoreline inhabited by mangroves. Ajaegwu *et al.* (2012) discussed the Late Miocene to Early Pliocene palynostratigraphy and palaeoenvironments of Ane-1 well, Eastern Niger Delta, Nigeria. They adopted the alpha-numerical method (Evamy *et al.*, 1978; Morley, 1997) to identify eight palynological zones, dated Late Miocene to Early Pliocene. Ojo and Adebayo (2012) carried out palynostratigraphy and palaeoecology of Chev-1 well, south-western Niger Delta basin. They identified nine palynozones and suggested that the studied sediments were deposited during Miocene-Pliocene period in which there was predominance of a high sea level and wet-humid climatic conditions because of the recovered palynomorphs were mainly made up of mangrove swamp floras. Osokpor *et al.* (2015) carried out palynozonation and lithofacies cycles of Paleogene to Neogene age sediments in PML-1 well, Northern Niger Delta Basin. They established two palynozones (*Ephedra claricristata* and *Auricupollenites echinatus* range zones) of Oligocene (Late Rupelian and Chattian stage) and three palynozones (*Verrutricolporites laevigatus/Verrutricolporites scabratus* range zone; and *Verrutricolporites rotundiporus* and *Margocolporites* sp. abundance zones) of Early–Late Miocene. The bases for the establishment of the zones are very vague. They neither followed properly the international stratigraphic guidelines nor the alpha-numeric methods of biozonation. Besides, they did not account for some intervals in between the zones. The aim of this work is to carry out the palynofacies and sedimentological studies of the strata penetrated by Ida-6 well in order to establish the palynostratigraphic zonation in line with the international stratigraphic guidelines as well as establish biochronology and palaeoenvironment of deposition of the strata penetrated by the well for the purpose of petroleum exploration.

6.1.1 Location of the studied well

The Niger Delta is located between latitudes 4° and 6°N and longitudes 3° and 9°E in Southern Nigeria. IDA-6 well is situated in the Ida oil field in the Niger delta. It is situated at 4.73°N and 6.96°E in the Coastal Swamp Depobelt of the Eastern Niger Delta (Figure 1).

6.1.2 Geology of the Niger Delta

The geology and stratigraphy of the Tertiary Niger Delta has been described by Short and Stauble (1967). They recognized three formations. In ascending order, these are the Akata, Agbada, and Benin formations (Figure 2). The Akata Formation generally consists of open marine and prodelta dark grey shale with lenses of siltstone and sandstone. Some sand beds considered to be of continental slope channel fill and turbidite are present (Weber and Daukoru, 1975). Thin sandstone lenses occur near the top, particularly near the contact with the overlying Agbada Formation. An estimated maximum thickness of the Akata Formation is possible only in the northern part of the delta where the formation has been drilled through into the Cretaceous

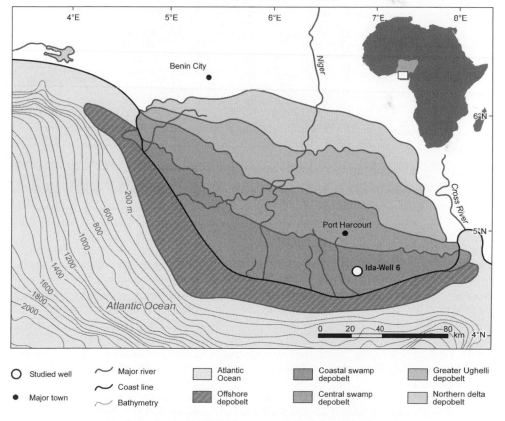

Figure 1. Depobelt map of the Niger Delta and location of the studied well (modified after Okosun and Chukwuma-Orji, 2016).

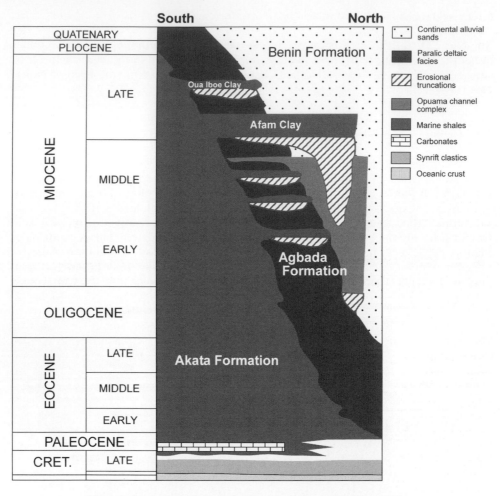

Figure 2. Stratigraphy of the formations of the Niger Delta (Ige, 2010).

(Avbovbo, 1978). A thickness range of 600 m to probably more than 6000 m is suggested by Weber and Daukoru (1975) and Durugbo and Uzodimma (2013). The age of the Akata Formation ranges from the Paleocene in the proximal parts of the delta to recent sediments in the distal offshore. The Agbada Formation consists of cyclic coarsening-upward regressive sequences. The coarsening upward sequences are composed of shales, siltstones, and sandstones which include delta front and lower delta plain deposits (Weber, 1971). The thickness of the Agbada sequences is highly variable (from 300 m up to about 4500 m). The oldest deposits of the Agbada Formation are of Eocene age in the north and are presently being deposited in the nearshore shelf domain. The Benin Formation comprises a succession of massive poorly indurated sandstones, thin shales, coals, and gravels of continental to upper delta plain origin. The Benin Formation first occurs in Oligocene times in the northern delta sector (Reijers *et al.*, 1996). The Benin Formation is up to 2000 m thick in the central onshore part of the delta and thins towards the delta margins (Bustin, 1988). Doust and Omatsola (1990) recognized depobelts in the Niger Delta which are distinguished primarily by their age and most importantly their location. These are the Northern Delta, Greater Ughelli, Central Swamp, Coastal Swamp, and Offshore depobelts (Figure 1). Each depobelt is filled with paralic sediments and bounded by faults at its proximal and distal limits. The paralic sedimentation in each depobelt resulted from eustatic sea level oscillations (transgression and regression or rise and fall) active in the basin within the development of the depobelt. During transgression, marine sediments are deposited, while during regression, continentally derived sediments are deposited (Ige, 2010; Bankole, 2010). The cyclicity of paralic sedimentation has been attributed to changes in climate associated with wet and dry climates (Burke and Durotoye, 1970; Bankole, 2010).

6.2 MATERIALS AND METHODS

The ditch cutting samples and wireline logs of Ida-6 well were provided by Chevron Nigeria Plc. Other materials required were made available by Crystal Age Limited, Lagos, Nigeria where sample processing and laboratory analyses were carried out.

For sedimentological analysis, the lithologic description of the stratigraphic intervals studied was based on the study of the log motifs (gamma ray and resistivity logs), physical inspection of the ditch cuttings, and microscopic study of the washed samples. Twenty grams of each sample was crushed and soaked with water and liquid detergent for 24 hours. The soaked samples were briskly washed under a distilled water nozzle tap using a 63 micrometre (μm) sieve mesh. The retained samples on 63 μm sieve were dried over hot plates and bagged for sedimentological studies. The bagged samples were spread on a black anodized aluminium foraminiferal picking tray and viewed using the standard binocular reflected light microscope (Fisher Scientific, No. 62416). The lithologic description was enhanced by the gamma-ray and resistivity logs since high and low values of gamma log and deep induction resistivity log signified shale and sand lithologies, respectively (Olayiwola and Bamford, 2016). The essential parameters studied were: (i) the rock types; (ii) colour and texture such as grain size, sorting and grain shape (roundness); and (iii) accessory mineral and fossil contents. The lithological data were plotted using Stratabug software to generate the vertical lithofacies profiles encountered within the studied intervals.

The standard acid palynological preparation method was followed. Fifty ditch cutting samples from Ida-6 well were subjected to analysis. Fifteen grams of each sample were treated with 10% HCl under a fume cupboard for the complete removal of carbonates. This was followed by neutralization with distilled water before the next procedure. Then 40% HF was added to the sample which was placed on a shaker for 24 hours to speed up the reaction and to ensure a complete dissolution of the silicates and for the particles to settle down. Thereafter, the HF was carefully decanted, followed by neutralisation with distilled water in order to remove fluoro-silicate compounds usually formed from the reaction with HF. Sieving and separation were performed using Brason Sonifier 250. Brason Sonifier is an electric device used with the aid of 5 micron sieve to filter away the remaining inorganic matter (silicates, clay, and mud) and heavy minerals to recover organic matters. The sieved residue was given controlled oxidation using concentrated nitric acid (HNO_3). The level of oxidation required by each sample was closely monitored under a microscope. The same procedure for sample preparation for palynomorphs recovery was followed for the palynomacerals, except that the oxidation process with HNO_3 was omitted in order not to bleach the palyno debris. The recovered organic matters were uniformly spotted on arranged cover slips of 22/32 mm and were then allowed to dry for mounting.

The mounting medium used for permanent mounting of cover slip onto glass slide was Loctite (Impruv) and was dried with natural sunlight for five minutes. The slides were then stained with safaranin-O in order to enhance the study of dinoflagellete cysts.

Both palynology and palynofacies slides were examined under the Olympus Binocular light transmitted microscope. The palynofacies slides were subjected to quantitative analysis of palynomacerals (Type 1, 2, 3, and 4) as well as structure-less organic matter (SOM). Identification of palynomorph and palynomacerals were done through the use of palynological albums and the published works of previous researchers (Germeraad et al., 1968; Ajaegwu et al., 2012; Bankole, 2010; Durugbo and Aroyewun, 2012; Ige, 2009; Ige et al., 2011).

6.3 RESULTS AND DISCUSSION

6.3.1 Lithologic description and sedimentological analyses

The lithology consists of alternating shale/mudstone and sandstone units with sandy shale and siltstone units (Figure 3). The shale/mudstones are mostly grey to brownish grey in colour, moderately hard to hard, platy to flaggy in appearance. The sandstones are white to very light gray, coarse to fine grained, angular to subangular to rounded, and poorly to well sorted in texture. The accessories include shell fragments, ferruginous materials, mica flakes, and carbonaceous detritus (Figure 3).

Sedimentological deductions were based on the integration of wire line log motifs, textural/lithologic attributes and the distribution of the accessory materials. These had permitted the assignment of the entire studied sections of Ida-6 well to the Agbada Formation. These criteria also enabled the recognition of one broad lithofacies sequence—The Paralic Lithofacies sequence within the sections (Figure 3). The Paralic Lithofacies Sequences were subdivided further into an Upper and Lower Paralic unit. The Upper Paralic unit consists dominantly of fine to medium grained, occasionally coarse to very coarse grained/granule sized, moderately to well sorted, subangular to subrounded sand grains with reddish brown to grey, platy to flaggy, moderately soft to moderately hard shales. The Lower Paralic unit consists of very fine to medium grained, occasionally coarse grained, well sorted and subangular sand grains with brownish grey, platy to flaggy, moderately soft to moderately hard shales.

6.3.1.1 Palynofacies

The charts in Figures 4 and 5 show the different palynomorph taxa and types of palynomacerals encountered at the different depths interval. Analysis of the slides of Ida-6 well yielded low to moderate palynomorph species. The diversity is also low to moderate. The palynomorphs recovered are pollens, spores and one indeterminate dinoflagellate cyst (Figure 4). The palynomacerals analysis yielded abundant records of palynomacerals 1 and 2, few occurrences of palynomacerals 3. There are no records of palynomacerals 4 and structure-less organic matter (SOM) (Figure 5). Microphotographs of some of the recovered palynomacerals and palynomorphs are presented on plates 1 and 2 respectively.

Palynomaceral 1 (PM 1)

The observed PM 1 are plant debris that appeared orange-brown to dark-brown in colour, opaque, irregular in shape, structure-less and varies in preservation (Plate 1). Batten and Stead (2005), and Oyede (1992) described PM 1 as particulate organic matter (Alginite) that is orange-brown to dark-brown in colour, dense in appearance,

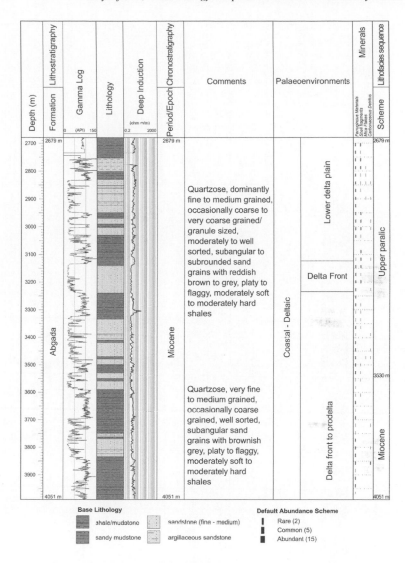

Figure 3. Lithological and sedimentological chart of Ida-6 well, Well code:
IDAMA-6-S, Interval: 2679–4051 m.

irregular in shape, structure-less, and varies in preservation. It is heterogeneous and of higher plant in origin and some are products of exudation processes such as the gelification of plant debris in the sediments. PM 1 includes small, medium and large sizes of flora debris, humic gel-like substances, and resinous cortex irregularly shaped materials (Batten and Stead, 2005; Oyede, 1992; Thomas *et al.*, 2015).

Palynomaceral 2 (PM 2)
The PM 2 are irregular in shape, brown-orange in colour, and platy-structured plant materials (Plate 1). PM 2 (Exinites) had been described to be usually brown–orange colour, structured, but irregular in shape. It encompasses platy like structured plant materials (leaves, stems or small rootlet debris), algae debris, and a few amounts of

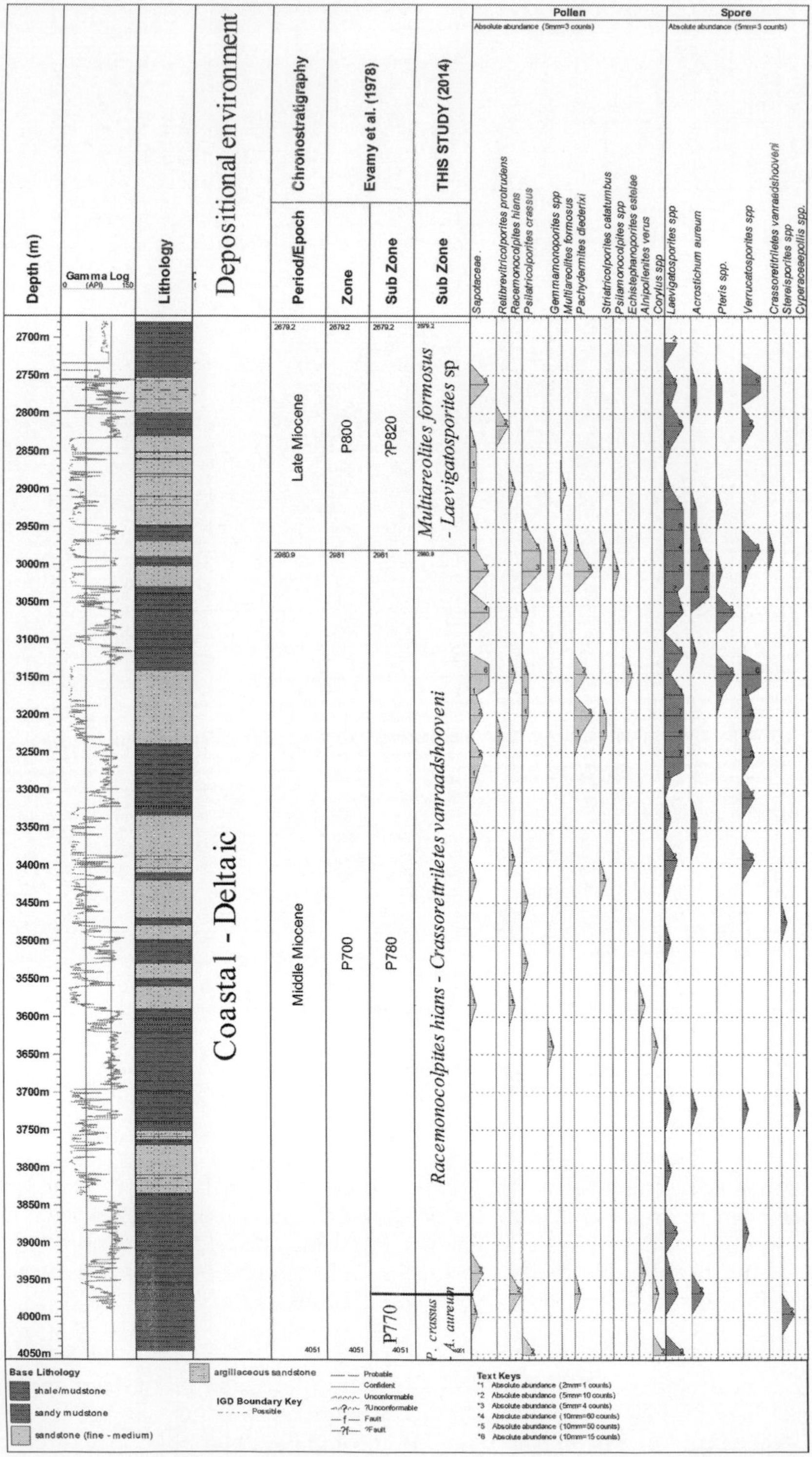

Depth (m)
Gamma Log
Lithology
Depositional environment
Chronostratigraphy
Period/Epoch
Evamy et al. (1978)
Zone
Sub Zone
THIS STUDY (2014)
Sub Zone
Pollen
Spore
Sapotaceae
Retibrevitricolporites protrudens
Racemonocolpites hians
Psilatricolporites crassus
Gemmamonoporites spp
Multiareolites formosus
Pachydermites diederixi
Striatricolporites catatumbus
Psilamonocolpites spp
Echistephanoporites estelae
Alnipollenites verus
Corylus spp
Laevigatosporites spp
Acrostichum aureum
Pteris spp.
Verrucatosporites spp
Crassoretitriletes vanraadshooveni
Stereisporites spp
Cyperaceaepollis spp.
Late Miocene
Middle Miocene
P800
P700
?P820
P780
P770
Coastal - Deltaic
Multiareolites formosus - Laevigatosporites sp
Racemonocolpites hians - Crassorettriletes vanraadshooveni
P. crassus - A. aureum
Base Lithology
shale/mudstone
sandy mudstone
sandstone (fine - medium)
argillaceous sandstone
IGD Boundary Key
Possible
Probable
Confident
Unconformable
?Unconformable
Fault
?Fault
Text Keys

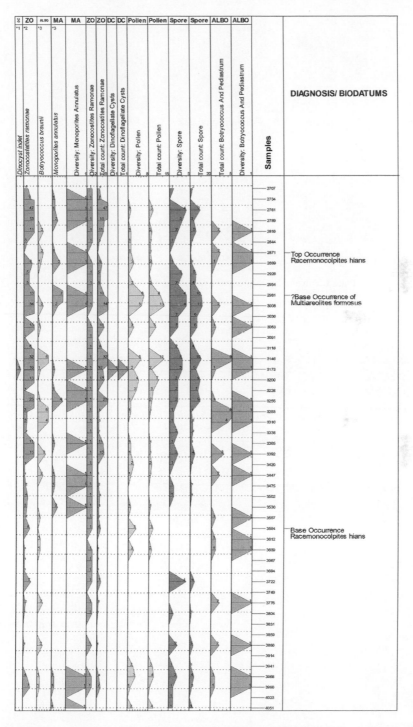

Figure 4. Palynomorph distribution chart of Ida-6 well; Well code: IDAMA-6-P, Interval: 2878–4051 m.

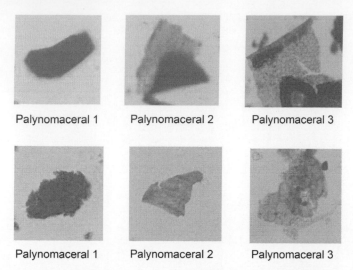

Palynomaceral 1   Palynomaceral 2   Palynomaceral 3

Palynomaceral 1   Palynomaceral 2   Palynomaceral 3

Plate 1. Palynomacerals recovered from the studied well (×400).

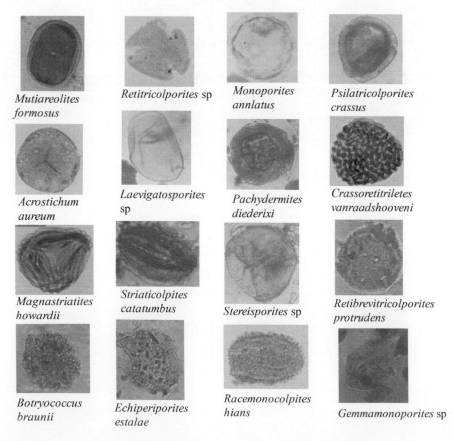

Mutiareolites formosus — *Retitricolporites* sp — *Monoporites annlatus* — *Psilatricolporites crassus*

Acrostichum aureum — *Laevigatosporites sp* — *Pachydermites diederixi* — *Crassoretitriletes vanraadshooveni*

Magnastriatites howardii — *Striaticolpites catatumbus* — *Stereisporites* sp — *Retibrevitricolporites protrudens*

Botryococcus braunii — *Echiperiporites estalae* — *Racemonocolpites hians* — *Gemmamonoporites* sp

Plate 2. Palynomorphs recovered from the studied well (×400).

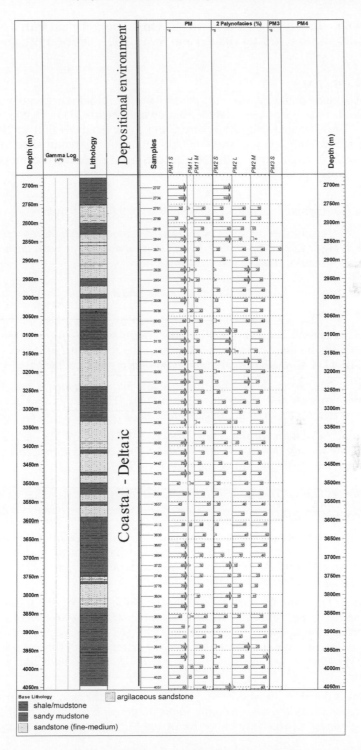

Figure 5. Palynomaceral distribution chart of Ida-6 well, Well code: IDAMA-6-P, Interval: 2679–4051 m.

humic gels and resinous substances. PM 2 is more buoyant than PM 1 because of its thinner lath-shaped character (Batten and Stead, 2005; Oyede, 1992; and Thomas *et al.*, 2015).

Palynomaceral 3 (PM 3)

PM 3 showed plant materials that are pale to brown in colour, irregular in shape, translucent and contain stomata (Plate 1). Batten and Stead (2005), Oyede (1992) and Thomas *et al.* (2015) stated that PM 3 (Vitrinite) is pale, relatively thin and irregularly shaped, and occasionally contains stomata. Also, it includes structured plant material, mainly of cuticular origin, and degraded aqueous plant material. It is more buoyant than PM 2 (Thomas *et al.*, 2015).

6.3.1.2 Palynostratigraphic zonations and biochronology of Ida-6 well

The biozones established are characterised by *Multiareolites formosus— Lavigatosporites* sp., *Racemonocolpites hians—Crassoretitriletes vanraadshoveni*, and *Psiltricolporites crassus—Acrostichum aureum* zones.

Multiareolites formosus–Lavigatosporites sp. Zone (Interval Zone)

Stratigraphic interval: 2707–2981 m

Definition: The interval between the FDO of *Lavigatosporites* sp. and the LDO of *Multiareolites formosus* defines the zone. The top of the zone is at 2707 m while the base is at 2981 m. LDO of *Crassoretitriletes vanraadshoveni* also marks the base of the zone.

Characteristics: The zone shows the presence of *Verrucatosporites* sp., *Pteris* sp., *Psiltricolporites crassus*, and *Racemonocolpites hians*. The zone is also characterised by moderate to rich occurrence of *Zonocostites ramonae* and *Laevigatosporites* sp.

Age: The zone is dated to the Late Miocene because of the association of *Crassoretitriletes vanraadshoveni, Multiareolites formosus*, and *Botryococcus brannii*, which are indicative of the Late Miocene (Morley, 1997)

Racemonocolpites hians–Crassoretitriletes vanraadshoveni Zone (Interval Zone)

Stratigraphic interval: 2981–3968 m

Definition: The top of the zone is marked by the LDO of *Crassoretitriletes vanraadshoveni* at 2981 m; while the base is marked the LDO of *Racemonocolpites hians* at 3968 m. FDO of *Pachydermites diederix, Striatricolporites catatumbus* and *Gemmamonoporites* sp. also characterise the top.

Characteristics: The zone is characterised by the association of the following taxa: *Racemonocolpites hians, Retibrevitricolporites protrudens, Pachydermites diederix, Striatricolporites catatumbus, Psilatricolporites crassus, Gemmamonoporites* sp., and Sapotaceae. The zone has moderate to rich occurrences of *Verucatosporites* sp., *Zonocoastites ramoneae, Laevigatosporites* sp., *Botryocococus brauni*, and a dinocysts indeterminate.

Age: The zone is dated to the Middle Miocene because of the association of the above mentioned taxa such as *Racemonocolpites hians, Retibrevitricolporites protrudens*, and *Pachydermites diederix*, which are indicative of the Middle Miocene (Morley, 1997).

Psilatricolporites crassus–Acrostichum aureum Zone (Interval Zone)

Stratigraphic interval: 3968–4051 m

Definition: The interval between the FDO of *Psilatricolporites crassus* and *Acrostichum aureum* at 4051 m and 3968 m respectively, defines the zone.

Characteristics: The zone is marked by the association and low occurrence of the following taxa: *Zonocoastites ramonae, Laevigatosporites* sp., *Botryococcus brauni, Coryius* sp. and *Acrostichum aureum*.

Age: The zone is dated to the Middle Miocene.

6.3.1.3 Palaeoenvironmental conditions of deposition

The Palaeoenvironment during the deposition involves the periodic changes in the depositional environment over geological time. Evaluation of the palaeoenvironment of deposition is essential because different depositional environments give rise to reservoirs with different qualities and characteristics such as porosity, permeability, heterogeneity, and architecture. Inference of the palaeodepositional environments of the studied wells was made based on the following criteria:

The palynoecological groupings of the recovered palynomorphs and the association of environmentally restricted diagnostic species such as *Zonocostites ramonae, Psilatricoloporites crasssus* (mangrove), *Monoporites annulatus* (montane), *Pachydermites diederixi* (fresh water swamp), *Laevigatosporites* sp., and *Botryococcus braunii* (rainforest) is shown in Table 1. Generally, the palynoecological groupings of the recovered palynomorph taxa indicate that the mangrove taxa have highest representation of the total recovery, followed by freshwater and rainforest swamps taxa in the well. Montane and savanna taxa have the lowest representation (Figure 6). Some authors (Adojoh *et al.*, 2015; Olayiwola and Bamford, 2016) agree that landward shifting of coastlines during sea level rise result in deposition of marine sediments in the subaerial delta plain. This period is also associated with shifting of the mangrove and other coastal swamp plant belts due to their preference for saline water. Therefore, the subaerial delta plain depositional environment is characterised by high representation of mangrove, other coastal swamp plants (from beach, brackish, freshwater swamp, rainforest, and palm) miospores, fungal elements, freshwater algae, and marine species (Adojoh *et al.*, 2015; Olayiwola and Bamford, 2016) (Figure 7). Similarly, during sea level fall, the coastline is shifted basinward and the shelf area ini-

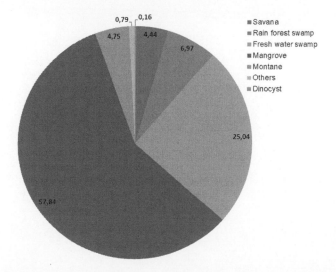

Figure 6. Palynoecological groups (%) of the recovered palynomorph taxa.

tially covered by marine water become exposed and probably incised due to erosion by fluvial activities. This results in deposition of terrestrial sediments in the subaqueous delta plain, which then become characterised by widespread savanna and montane vegetation belts (Figure 7). This depositional environment is characterised by maxima spectra of savanna and montane pollen (Adojoh *et al.*, 2015; Olayiwola and Bamford, 2016; see also Germeraad *et al.*, 1968; Jennifer *et al.*, 2012; Olajide *et al.*, 2012).

The type/nature of organic matter (palynomercerals) recovered from the samples point towards terrestrial/coastal and marine depositional environments, which have

Table 1. Palynoecological groupings of palynomorph taxa recovered from IDA-6 well.

S/N	Sample depth (meters)/taxa	Savanna Taxa							Montane Taxa			RFS Taxa			Freshwater swamp Taxa									Mangrove Taxa				Others			Dinocyst
		Pteris sp.	*Cyperaceapollis* spp	*Corylus* spp	*Stereisporites* sp.	Fungal spore	*Retibrevitricolporites protrudens*	Total savana taxa	*Monoporites annulatus*	*Ainipollenites verus*	Total Montane taxa	*Pachydermites diederxi*	*Sapotaceae*	Total RainForest Swamp taxa	*Striatricolporites catatumbus*	*Gemmamonoporites* sp.	*Racemonocolpites hians*	*Multiareolites formosua*	*Crassoretitriletes vanraadshooveni*	*Laevigatosporites* sp.	*Botryococcus braunii*	*Verrucatosporites* sp.	Total Freshwater Swamp taxa	*Zonocostatites ramonae*	*Psilatricolporites crassus*	*Acrostichum aureum*	Total mangrove taxa	*Psilamonocolpites* sp.	*Echistephanoporites estalae*	Total others	dinocyst indeterminate
1	2707							0			0			0						2			2	4			4			0	
2	2734							0	1		1			0									0	7			7			0	
3	2761	1						1	1		1	3		3						2		5	7	47		1	48			0	
4	2789	1				1		2	2		2			0						1			1	13		1	14			0	
5	2816					2		2			0			0						3	2	2	7	11			11			0	
6	2844							0			0		1	1						1			1	3			3			0	
7	2871							0	1		1		1	1							2		2	3			3			0	
8	2899							0	3		3		1	1				1		1	1		3	8			8			0	
9	2926	1						1			0			0						3			3	5		1	6			0	
10	2954							0			0		1	1						3			3	5	1	1	7			0	
11	2981							0	2		2	1	1	2	1	1		1	1	4		3	11	13	3	2	18			0	
12	3008	1						4	2		2	3		3		1				5	2	2	10	34	3	4	41		1	1	
13	3036							0			0			0						2			2			4	4			0	
14	3063	3						3			0		4	4						3	1		4	13	1		13			0	
15	3091							0			0			0									0	1			1			0	
16	3118							0			0			0						8			8	32		1	33			0	
17	3146	3			4	2		9			0	2	6	8	1					5	5	1	12	8	1		9	1		1	
18	3173	1		3				4	7		7	1		1						3	1	1	5	10	1		10			0	1
19	3200	4						4	1		1	3	2	5	1					4		2	7	13	1		14			0	
20	3228			4		1		5	4		4	1		1	1					6	1		8	1			1			0	
21	3255							0	3		3	2		2						7	1	2	10	23			23			0	
22	3283							0			0	1		1						1	6		7	3			3			0	
23	3310				1			1			0			0							4	2	6	3			3			0	
24	3338							0			0			0						1			1	2		1	3			0	
25	3365							0	1		1	1		1									0	11	1		12			0	
26	3392							0			0			0					1	2	3	2	8	13			13			0	
27	3420				1			1			0	1		1	1					1			2	1			1			0	
28	3447							0	1		1			0						2			2	1	1		2			0	
29	3475			1				1	1		1			0									0	2			2			0	
30	3502							0			0			0						1			1	4			4			0	
31	3530							0	2		2			0									0	3	1		4			0	
32	3557							0			0			0								1	1	2			2			0	
33	3584							0		1	1	1		1				1					1	6			6			0	
34	3612							0			0			0						1			1				0			0	
35	3639		1					1			0			0									0	1			1			0	
36	3667							0			0			0									0	1			1			0	
37	3694							0			0			0									0	1			1			0	
38	3722		1					1			0			0						1			1	7		1	8			0	
39	3749							0			0			0									0				0			0	
40	3776							0			0			0						1			1	2			2			0	
41	3804							0			0			0						1			1	1			1			0	
42	3831							0			0			0									0				0			0	
43	3859							0			0			0									0				0			0	
44	3886							0			0			0						2	2	1	5	1			1			0	
45	3914							0			0			0									0				0			0	
46	3941							0		1	1	2		2						1			1				0			0	
47	3968		1					1	1		1	1		1				2		2	1		5	4		2	6			0	
48	3996			2				2	1		1	1		1						1			1	7			7			0	
49	4023							0			0			0									0			1	1			0	
50	4051			1				1	1		1			0						3			3				0	2		2	

RFS = Rain forest Swamp

been distinguished to have distinctive and characteristic palynofacies (Oyede, 1992; Thomas *et al.*, 2015). The terrestrial/coastal environments are characterised by poorly sorted PM 1 and 2, absence of dinocysts, and common to abundant occurrence of fungal spores; while marine environment is characterised by a good sorting of organic matter predominantly small to medium, common to abundant PM 1 and 2, some needle-shaped to lath-shaped PM 4, and presence of dinocysts and/or foraminifera linings (Oyede, 1992). The wire line log motifs, lithology, and accessory minerals were also considered in the palaeoenvironmental determination.

Based on these criteria, the lower delta plain to delta front and prodelta (subaerial delta to subaqueous delta plains) environments within coastal-deltaic environment of deposition have been inferred for the sediments encountered in the analysed intervals of Ida 6 well (Table 2, Figure 3).

The interval 2679–3135 m in Ida-6 well was delineated to have been deposited in the lower delta plain environment (Figure 3). The interval is characterised by high occurrences of mangrove taxa (*Zonocostites ramonae–Rhizophora, Psilatricoloporites crasssus* and *Acrostichum aureum*), followed by freshwater swamp (*Laevigatosporites* sp., and *Botryococcus brannii*) and rainforest swamp taxa (sapotaceae and *Pachydermites diederixi*). This interval has also little representation of savanna (*Retibrevitricolporites protudens, Pteris* sp., and fungal spore) and montane taxa (*Monoporites*

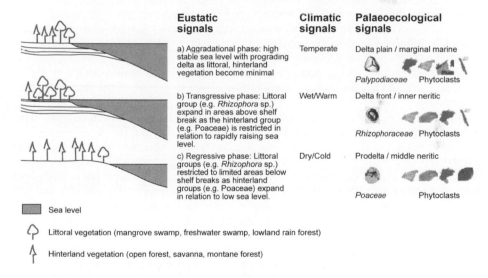

Figure 7. Relationship between palaeoecology, palaeovegetation, eustasy and climate in the tropical setting (modified after Adojoh *et al.*, 2015).

Table 2. Environment of deposition in Ida-6 well.

Ida 6 well interval (m)	Inferred depositional environment
2679–3135	Subaerial delta (lower delta plain)
3135–3240	Subaqueous delta (delta front) plain
3240–4051	Subaqueous delta (delta front to prodelta) plain

annulatus—Poacaea) (Table 1). High representation of coastal miospores (mangrove, fresh water and rainforest taxa) compared to minimal representation of hinterland miospores (savanna and montane taxa) are characteristics of subaerial sediments that were deposited during sea level rise and wetter climate; while the opposite represents sea level fall and drier climate (Morley, 1995; Ige, 2009; Adojoh *et al.*, 2015; Olayiwola and Bamford, 2016). The higher occurrences of mangrove, rainforest and fresh water taxa within the interval are indications that the interval was deposited in lower delta plain during sea level rise and wet climate (Figure 7). The prevailing climatic conditions supported the flourishing of the mangrove, rainforest, and fresh water vegetations. This deduction agrees with that of some previous researchers in the Niger Delta (Durugbo *et al.*, 2010; Ige *et al.*, 2011; Ola and Adewale, 2014; Bankole *et al.*, 2014). They utilised high percentage occurrences of mangrove taxa (Rhizophora), fresh water and rainforest taxa to delineate wet climatic zones, which were also an indication of rise in sea level. The abundant records of poorly sorted palynomacerals 1 and 2 (Figure 5) indicate coastal deltaic environment of deposition with influx of fresh water from moderate quantities of *Botryococcus braunii* and *Laevigatosporites* sp. recorded within the interval (Oyede, 1992). The recorded PM 1 and 2 consist of large, medium and small sizes.

Aggradational, progradational and retrogradational log motifs characterise the sandstone units (intercalated by shales and silts) in the intervals suggest their deposition as channel/bar complexes in a delta plain–delta front setting (Selley, 1978). Lithologically, the sand grains are white, very fine to medium–grained, occasionally coarse to very coarse-grained/granule-sized, poorly to well sorted, and subangular to subrounded. The shales are reddish brown to grey, platy, and moderately soft to moderately hard. The accessory mineral suites are mostly ferruginous materials, shell fragments, carbonaceous detritus, and mica flakes in decreasing order of abundance. These criteria indicate deposition in lower deltaic plain (inner neritic) environments.

Similarly, the interval 3135–3240 m was delineated to have been deposited in delta front (inner neritic) environment of deposition (see Figure 3).

The criteria for this deduction are: The interval is characterised by increased occurrence of savanna taxa such as *Pteris* sp. and fungal spores, and *Monoporite annulatus* (montane taxa). The reduced occurrences of mangrove, freshwater swamp, and rainforest swamp taxa compared to the above interval indicates dry climate and sea level fall and sediments deposition in delta front environment (Adojoh *et al.,* 2015). Durugbo *et al.* (2010), Ige (2009) and Bankole *et al.* (2014) have utilized this approach of low representation of mangrove taxa (Rhizophora), fresh water and rainforest taxa, and high representation savanna and montane taxa to delineate dry climatic zones.

The PM 1 and 2 that occur are more of large and medium size than small size (Figure 5).

This interval is characterised by blocky/aggradational log motifs (slightly serrate cylinder on funnel–shaped log character), suggesting deposition as channels fills in a delta front setting. Accessory minerals are dominated by ferruginous materials and shell fragments. Carbonaceous detritus showed spotty occurrences and absence of glauconite pellets may suggest a distributary channel-subaqeous mouth bar.

Finally, the interval 3240–4051 m was inferred to have been deposited in delta front to prodelta (inner to middle neritic) environment of deposition (see Figure 3).

The reasons for this inference are: The interval is characterised by moderate occurrence of mangrove taxa (*Zonoccostatites ramonae*, *Psilaticolporites crassus* and *Acrostichum aureum*), freshwater taxa (*Striatricolporites catatumbus*,

Levigatosporites sp., *Botryococcus brauni*, *Verrucatosporites* sp. and *Gemmamonoporites* sp.), and rainforest taxa (*Pachydermites diederixi* and sapotacea). This is supported by rare to no presence of savanna (*Coryius* sp., *Cyperaceapollis* sp., *Pteris* sp., and fungal spores) and montane taxa (*Monoporites annulatus*) (Table 1). These suggest sediment deposition in subaqueous delta during stable and temperate climatic conditions and a relative decrease in sea level (Figure 7).

The interval is also characterised by moderate to good sorting of palynomacerals 1 and 2, predominantly common to abundant small to medium sizes.

The predominantly shaly/silty character of the lower section; and the presence of ferruginous materials, shell fragments and carbonaceous detritus suggest deposition in low energy, oxic, shallow marine settings (Selley, 1978). The sand units exhibited multiserrate funnel, cylinder/subtle bell-shaped gamma ray log profiles interpreted as subaqueous mouth bars and distributary channel deposits indicate a prograding shoreline.

6.4 CONCLUSION

Palynofacies and sedimentological analyses were carried out on the strata penetrated by Ida-6 well using the ditch cuttings and wire line logs data provided by Chevron Nigeria Plc. The recovery of the palynomorphs was not very rich and hence poorly diversified. However, three palynostratigraphic zones (interval zones) have been established using the international stratigraphic guide for establishment of biozones (Murphy and Salvador, 1999). They are *Multiareolites formosus–Lavigatosporites* sp., *Racemonocolpites hians—Crassoretitriletes vanraadshoveni,* and *Psiltricolporites crassus—Acrostichum aureum* zones. The studied stratigraphic interval was dated to the Middle Miocene to Late Miocene base on the recovered Miocene age diagnostic marker species such as *Multiareolites formosus, Verrutricolporites rotundiporus, Crassoretitriletes vanraadshoveni,* and *Racemonocolpites hians*. The lithology showed alternation of shale and sandstone units with few intercalations of siltstone units. Accessories minerals are dominated by ferruginous material, shell fragments, and carbonaceous detritus with spotty occurrence of mica flakes. The lithologic, textural, and wire line log data indicate that the entire studied interval belong to the Agbada Formation. Coastal-deltaic (lower delta plain to prodelta) environments of deposition have been inferred for the studied interval on the bases of the palynofacies association and sedimentological characteristics. The variations in the relative abundance of the recovered palynomorphs (hinterland versus coastal/lithoral) taxa are characteristics of different palaeoenvironments of prograding paralic succession (Morley, 1995). The higher occurrences of mangrove, rainforest, and fresh water taxa (lithoral/coastal vegetation) with few savanna and montane taxa within the interval (2679–3135 m) are indications that the interval was deposited in lower delta plain during sea level rise and wet climate (Figure 7). The increased savanna and montane taxa within 3135–3240 m of the studied interval and reduced occurrences of mangrove, freshwater swamp, and rainforest swamp taxa compared to the interval above indicate dry climate, sea level fall, and sediment deposition in a delta front environment (Adojoh *et al.*, 2015). The moderate occurrence of mangrove, rainforest, and fresh water taxa rare to no presence of savanna and montane taxa within 3240–4051 m suggest sediment deposition in a subaqueous delta (delta front to prodelta) during stable and temperate climatic conditions and a relative decrease in sea level.

REFERENCES

Adojoh, O., Lucas F.A. and Dada, S., 2015, Palynocycles, Palaeoecology and Systems Tracts Concepts: A Case Study from the Miocene Okan-1 Well, Niger Delta Basin, Nigeria. *Applied Ecology and Environmental Sciences*, **3**, pp. 66–74. doi: 10.12691/aees-3-3-1.

Ajaegwu, N.E., Odoh, B.I., Akpunonu, E.O., Obiadi, I.I. and Anakwuba, E.K., 2012, Late Miocene to Early Pliocene Palynostratigraphy and Palaeoenvironments of ANE-1 Well, Eastern Niger Delta. *Nigeria Journal of Mining and Geology*, **48**, pp. 31–43.

Avbovbo, A.A., 1978, Tertiary lithostratigraphy of Niger Delta. *American Association of Petroleum Geologists Bulletin*, **62**, pp. 295–300.

Bankole, S.I., 2010, *Palynology and stratigraphy of three deep wells in the Neogene Agbada Formation, Niger Delta, Nigeria. Implications for petroleum exploration and paleoecology*, PhD thesis, Technische Universität Berlin, pp. 1–190.

Batten, D.J. and Stead, D.T., 2005, Palynofacies analysis and its stratigraphic application. In *Applied stratigraphy*, edited by Koutsoukos, E.A.M., (Dordrecht: Springer), pp. 203–226.

Burke, K. and Durotoye, B., 1970, Late Quaternary climate variation in the southwestern Nigeria, evidence from pediments and pediment deposits. *Bulletin d'Asequa*, **25**, pp. 79–96.

Bustin, R.M., 1988, Sedimentology and characteristics of dispersed organic matter in Tertiary Niger Delta: Origin of source rocks in a deltaic environment. *American Association of Petroleum Geologists, Bulletin*, **72**, pp. 277–298.

Doust, H. and Omatsola, E., 1990, Niger Delta Divergent/Passive Margin Basins. *American Association of Petroleum Geologists, Memoir*, **48**, pp. 201–238.

Durugbo, E.U. and Aroyewun, R.F., 2012, Palynology and Paleoenvironments of the Upper Araromi Formation, Dahomey Basin, Nigeria. *Asian Journal of Earth Sciences*, **5**, pp. 50–62. doi: 10.3923/ajes.2012.50.62.

Durugbo, E.U. and Uzodimma, E., 2013, Effects of lithology on palynomorph abundance in wells X1 and X2 from the Western Niger Delta, Nigeria. *International Journal of Geology, Earth and Environmental Sciences*, **3**, pp. 170–179.

Evamy, B.D., Haremboure, J., Karmerling, P., Knaap, W.A., Molloy, F.A. and Rowlands, P.H., 1978, Hydrocarbon habitat of the Tertiary Niger Delta. *American Association of Petroleum Geologists, Bulletin*, **62**, pp. 1–39.

Germeraad, J.J., Hopping, G.A. and Muller, J., 1968, Palynology of Tertiary sediments from tropical areas. *Review of Palaeobotany and Palynology*, **6**, pp. 189–348. doi: 10.1016/003667(68)90051-1.

Ige, O.E., 2009, A Late Tertiary Pollen record from Niger Delta, Nigeria. *International Journal of Botany*, **5**, pp. 203–215. doi: 10.3923/ijb.2009.203.215.

Ige, O.E., Datta, K., Sahai, K. and Rawat, K.K., 2011, Palynological Studies of Sediments from North Chioma-3 Well, Niger Delta and its Palaeoenvironmental Interpretations. *American Journal of Applied Sciences*, **8**, pp. 1249–1257.

Jennifer, Y.E., Mebradu, S.M., Okiotor M.E. and Imasuen, O.I., 2012, Palaeo-environmental studies of well "AX", in the Niger Delta. *International Research Journal of Geology and Mining*, **2**, pp. 113–121.

Morley, R.J., 1997, Offshore Niger Delta palynological zonation, prepared for the Niger Delta Stratigraphic Commission. *Palynova*, **1**, pp. 1–6.

Murphy, M.A. and Salvador, A., 1999, International Stratigraphic Guide—An abridged version, International Subcommission on Stratigraphic Classification of IUGS, International Commission on Stratigraphy, *Special Episodes*, **22**, 4, pp. 255–272.

Okosun, E.A. and Chukwuma-Orji, J.N., 2016, Planktic Foraminiferal Biostratigraphy and Biochronology of KK-1 Well Western Niger Delta, Nigeria. *Journal of Basic and Applied Research International*, **17**, pp. 218–226.

Olajide, F.A., Akpo, E.O. and Adeyinka, O.A., 2012, Palynology of Bog-1 Well, Southeastern Niger Delta Basin, Nigeria. *International Journal of Science and Technology*, **2**, pp. 214–222.

Olayiwola, M.A. and Bamford, M.K., 2016, Petroleum of the Deep: Palynological proxies for palaeoenvironment of deep offshore upper Miocene-Pliocene sediments from Niger Delta, Nigeria. *Palaeontologia Africana*, **50**, pp. 31–47.

Ojo, A.O. and Adebayo, O.F., 2012, Palynostratigraphy and paleoecology of chev-1 well, southwestern Niger delta basin, Nigeria. *Elixir Geoscience*, **43**, pp. 6982–6986.

Osokpor, J., Lucas, F.A., Osokpor, O.J., Overare, B., Elijah, I.O. and Avwenagha, O.E., 2015, Palynozonation and lithofacies cycles of paleogene to Neogene age sediments in PML-1 well, Northern Niger Delta Basin. *The Pacific Journal of Science and Technology*, **16**, pp. 286–297.

Oyede, A.C., 1992, Palynofacies in deltaic stratigraphy. *Nigerian Association of Petroleum Explorationist Bulletin*, **7**, pp. 10–16.

Reijers, T.J.A., Petters, S.W. and Nwajide, C.S., 1997, The Niger Delta Basin, sedimentary geology and sequence stratigraphy. In *African Basins*, edited by Selley, R.C., (Amsterdam: Elsevier), pp. 151–172.

Selley, R.C., 1978, Concept and methods of subsurface facies analysis. *American Association of Petroleum Geologists, Bulletin, Education course notes*, **9**, pp. 1–86.

Short, K.C. and Stauble, A.J., 1967, Outline of the geology of Niger Delta. *American Association of Petroleum Geologists, Bulletin*, **51**, pp. 761–779.

Thomas, M.L., Pocknall, D.T., Warney, S., Bentley, S.J. Sr., Droxler, A.W. and Nittrouer, C.A., 2015, Assessing palaeobathymetry and sedimentation rates using palynomarceral analysis: a study of modern sediments from the Gulf of Papua, offshore Papua New Guinea. *Palynology*, **39**, 3 pp. 1–24. doi/abs//10.1080/0191612 2.2015.1014526.

Weber, K.J., 1971, Sedimentological aspects of oil fields in the Niger Delta, *Geologie en Mijnbouw*, **50**, pp. 559–576.

Weber, K.J. and Dankoru. E.M., 1975, Petroleum geology of Niger Delta. In *Proceedings of the 9th annual conference on world petroleum congress*, (Tokyo: London Applied publishers' Ltd), **2,** pp. 209–221.

CHAPTER 7

Tropical palaeovegetation dynamics, environmental and climate change impact from the low latitude coastal offshore margin, Niger Delta, Gulf of Guinea

Onema Adojoh
School of Environmental Sciences, University of Liverpool, UK
Geosciences and Geological and Petroleum Engineering,
University of Missouri Science and Technology, Rolla, USA

Fabienne Marret & Robert Duller
School of Environmental Sciences, University of Liverpool, UK

Peter Osterloff
Shell UK Limited, Aberdeen, UK

ABSTRACT: The Niger Delta is located in a region of great sensitivity to climate change, offering the potential to provide an insight into environmental change during the Late Quaternary in Equatorial Africa. This deltaic environment and dynamically linked vegetation system had been sparsely studied if compared with other major tropical deltas and its response to past climate change and vegetation development is therefore poorly understood. This study provides a chronological interpretation of palynomorph data from three gravity cores that has enabled the differentiation of warm and dry climate intervals, and the factors controlling vegetation change (Intertropical Convergence Zone and West African Monsoon). An array of palynological sequences defined in the three gravity cores show very similar fluctuations, with a dominance of Afromontane Forest (Podocarpaceae), Freshwater Swamp (Cyperaceae), Savannah (Poaceae), and Lowland Rainforest (Polypodiaceae) elements during the late glacial (MIS2) and deglaciation (MIS1) periods, followed by the significant establishment of coastal fringing mangroves (Rhizophoraceae) during the Early to Mid-Holocene. In addition, high values of significance charred grass cuticles were observed in samples from the late glacial period, whereas the expansion of mangrove vegetation is concomitant with the relatively warmer interglacial period. These records suggest dry conditions during the late glacial and deglaciation periods, and warmer conditions during the subsequent interglacial period of the Early to Mid-Holocene periods, respectively. The data presented here demonstrate a close relationship between prevailing climate change and the control on vegetation evolution in the Niger Delta when compared with previous studies from Equatorial West Africa.

7.1 INTRODUCTION—ENVIRONMENTAL AND LOCATION SETTING

Delta systems are the interface between rivers draining the land and the ocean, and sites of long-term sediment accumulation which preserve pollen and organic matter storage that can be used for environmental investigations. Deltas are particularly sensitive to climate and sea level change, which are expressed as changes in the shelf and coastline morphology which influence associated vegetation composition and distribution patterns.

The Cenozoic Niger Delta has been noted as one of the largest delta systems in the world with an areal surface extent of about 300,000 km^2 (Kulke, 1995) and a postulated sediment thickness of over 12,000 m in the basin depocentre (Doust and Omatsola, 1990; Reijers, 2011). The delta system is axially positioned across southern Nigeria prograding roughly southwards into the Gulf of Guinea (Figure 1).

In terms of its vegetation distribution and environmental condition, the Niger Delta can be categorised along with other low latitude tropical environments such as South America, West and East Africa, South-East Asia and Indonesia. They are amongst the richest and most diverse ecological systems in the World with respect to their vegetation biomes. However, the mechanisms that control their evolution and spatial development over the long term are not fully understood (Morley, 2000; Dalibard *et al.*, 2014). This knowledge is essential if we are to forecast the future development of tropical vegetation through time. In particular, additional research is required to tie successive vegetation variations with key factors associated with climate change (e.g. Tjallingii *et al.*, 2008; Clemens *et al.*, 2010). To advance knowledge and predictive capacity in this area, a high-resolution temporal framework linked to

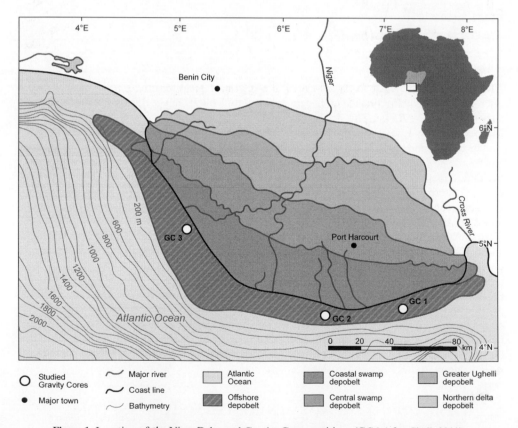

Figure 1. Location of the Niger Delta and Gravity Cores positions (GCs) (after Shell, 2011).

vegetation and ecological systems development is required to provide the framework for undertaking reconstructions of the past evolution of vegetation patterns. Interpretaton of such vegetation patterns will form part of an important contribution to environmental reconstruction models, which take vegetation/climate forcing into consideration. In most cases the vegetation/climate forcing relationship is directly linked to seasonal changes in the prevailing atmospheric pressure circulation (Gehrels, 1999; Dalibard *et al.*, 2014).

Seasonal variation in the position of the Intertropical Convergence Zone and West African Monsoon system (ITCZ and WAM), influenced by the interaction of atmospheric circulation and oceanic conditions, has influenced the prevailing climate and vegetation types through time and, therefore, seasons (Figure 2). The overall atmospheric circulation of the southern hemisphere is usually linked to the intensity and location of the easterly wind belt at tropical latitudes, the westerly wind belt at middle latitudes and the polar easterlies at polar latitudes (Leroux, 1993).

Across the Niger Delta, the understanding of this interplay of winds associated with atmospheric movement of the southern hemisphere circulation and its influence on vegetation dynamics are limited by the relatively small numbers of well-dated palynomorph records from the Gulf of Guinea, particularly on the coast of the Niger Delta. The Niger Delta has not received much extensive research in the fields of climate and vegetation evolution over the years due to restricted access to oil company records. Similarly, with other regions of West Africa, the understanding of palaeo-environmental change is also poorly developed, primarily due to the scarcity of stratified lakes or reducing environments provide favourable conditions for the preservation of palynomorph material (Massuanganhe *et al.*, 2016).

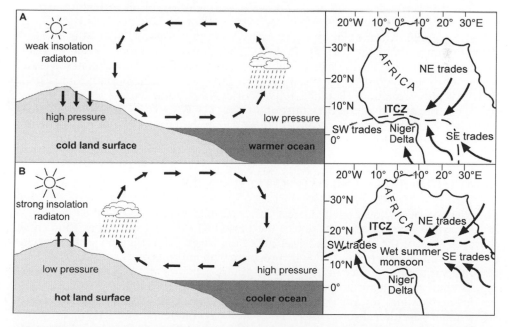

Figure 2. Seasonal positions of the Intertropical Convergence Zone (ITCZ) and West African Monsoon System (WAM) during the boreal winter (A) and summer (B). ITCZ is indicated by dotted lines (modified from Griffiths (1972) and Leroux (1993)).

To improve our knowledge of the evolution of vegetation communities and spatial patterns, predominant climate, sea level change impact, and littoral marine or nearshore marine conditions of the Niger Delta during the Late Pleistocene to the Holocene, detailed palynomorph data records were assessed. Three Gravity Cores (GCs) from the eastern (GC1), central (GC2), and western parts (GC3) of the Niger Delta (Nigeria) were selected and investigated (Figure 1). The palynomorph record from the deltaic and coastal boundary of the terrestrial and marine domains of the Niger Delta were analysed to ascertain the region's contributions towards understanding global and local climate change and the influence on vegetation dynamics of cyclical wet and dry conditions in the catchment.

Past climate records from other West African regions focused on key climate drivers have shown that climate and vegetation along the active deltaic systems in the tropics today are influenced by several atmospheric-oceanic interactions (e.g. Dupont and Agwu, 1991; Dupont *et al.*, 1998; Kim *et al.*, 2010; Dupont, 2011; Marret *et al.*, 2013). Therefore, in order to assess how the theory of the regional scale climate and vegetation components is affected by atmospheric and oceanic interactions, this study evaluates the hypotheses surrounding the relative position of the Intertropical Convergence Zone (ITCZ) and the intensity of the West African Monsoon (WAM) in relation to dynamic vegetation and climate change (Shanahan *et al.*, 2007) (Figure 2).

This study concerning the Niger Delta tests the hypotheses of the ITCZ and WAM migration during the Late Quaternary and the response of the Niger Delta vegetation with climate and sea level oscillations relative to other studies from West Africa. This paper highlights hinterland (terrestrial) and littoral vegetation (coastal marine) considered paramount to evaluate seasonality of the West African climate and sea level oscillation through interpretation of the stages in vegetation changes. The hinterland (terrestrial) vegetation components are indicative of all vegetation in arid regions where their existence is influenced by prevailing dry climates (e.g. some Open Forest; Savannah/Afromontane vegetation) (Adojoh and Osterloff, 2010; Marret *et al.*, 2013; Bouimetarhan *et al.*, 2015) (Table 2). The littoral vegetation (coastal marine) components represent all the vegetation influenced by sea level or wet climates in close proximity to the lower delta plain setting (e.g. Mangrove-Coastal Swamp; some Fresh Water Swamp vegetation) (Poumot, 1989; Morley, 1995; Rull, 2002).

This research explores the impact of the ITCZ and the WAM circulation dynamics through vegetation reconstructions within the Niger Delta during the Late Pleistocene—Mid-Holocene.

7.2 METHODOLOGY AND DATING TECHNIQUES

The selection of shallow Gravity Cores (GCs) was made on the basis of their positions in an East to West transect to reflect the presumed shoreline palaeoenvironmental characteristics of the littoral coastal setting of the Niger Delta (Figure 1). The selected GCs were anticipated to reflect deltaic processes with imprints of fluvial processes and seasonal variability, relative to tidal and longshore activities and overposting of those signals upon any structurally inherent signals. In addition to recording the marine versus fluvial dynamics, the analysis of pollen recovered from the GCs should encapsulate any climate variation through time. The palaeoenvironmental limits defined by the cores were expected to be between the Limit of Tidal Influence (LoTI) and Limit of Freshwater Influence (LoFI). However, other processes that influence the sediments and associated pollen records were also considered.

Table 1. GCs coordinates and core details.

Cores	GC1	GC2	GC3
Latitude	4°49′43″ N	4°05′08″ N	4°11′59″ N
Longitude	5°20′29″ E	6°33′30″ E	7°21′29″ E
Water depth	40 m	40 m	40 m
Length	272 cm	266 cm	260 cm

Three gravity cores (GCs) achieved by The Shell Petroleum Development Company of Nigeria (SPDC) were drilled by CGG Robertson Fugro in 2002 from the sea-bed within the shallow marine realm, in approximately 40 m water depth in the nearshore of the Niger Delta (Figure 1 and Table 1). The distribution of the sites selected was to enable recognition of local changes in analytical proxies, and link them more regionally across the Niger Delta from West to East. For each GC, approximately 3 m of core was collected and sampled at 2 cm intervals for detailed palaeoenvironmental investigation and vegetation reconstructions. Each core had been pristinely preserved and was to a certain extent "water wet" on being accessed for sampling, demonstrating the integrity of the core housing that kept the sediments in place/"right way up".

Applied research techniques were based on standard preparatory and descriptive methods for core lithological description and associated palynological analysis. The palynological components for this study were analysed following the laboratory routines suggested for the preparation of Low Latitude African samples (e.g. Dupont *et al.*, 2000; Marret *et al.*, 2001; Scourse *et al.*, 2005). Prepared samples were analysed using standard acid, reagent and sieving methodologies to establish the variety and abundance of palynological components (sporomorphs, palynodebris, non-palynomorphs, etc.). The strewn mount slides were examined using a Nikon microscope, and images of key specimens were captured. Where possible, a minimum of 300 palynomorphs and 120–200 mangrove taxa were counted using 400X magnification for the quantitative analysis. Finally, two statistical palynological models were adopted for the interpretation of pollen data, namely, the Palaeovegetational and Tidal Limit indexes. The former is the ratio between the scores of two major principal groups representing different vegetation types (Rull, 2002). These were altered in the present study to reflect littoral and hinterland vegetation changes influenced by climate and variations in sea level. To establish these indices and zonation, Tilia Version 1.7 and CONISS (Grimm, 1987) were used to plot possible vegetation groups/family affinities.

7.2.1 Chronology

Due to lack of radiocarbon-datable material such as shells or foraminifera, a refined biostratigraphic subdivision was developed based on nannofossil (NN) indicators obtained from a sub-set of 10 key samples. This subdivision defining the spatial and temporal evolution of the coastal vegetation of the Niger Delta was based on the first occurrence (FOC) of *Gephyrocapsa oceanica* (NN19) and *Emiliania huxleyi* (NN20/21) occurrences in the three GCs. This interpretation established a stratigraphical framework reflecting the Late Pleistocene to Mid-Holocene (Martini, 1971; Raffia *et al.*, 2006).

The three GCs presented similar age and depth trends that support generic ages of Late Pleistocene (~20 ka) from 275–184 cm depth to Mid-Holocene (~6.5 ka) at 184–

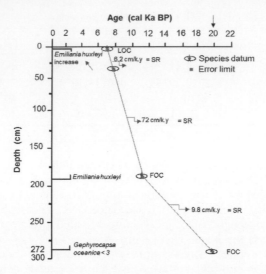

Figure 3.1. Sedimentation Rate (SR) and age-depth model for GC1.
Note: FOC = first occurrence in core, LOC = last occurrence in core.

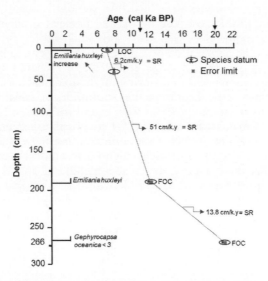

Figure 3.2. Sedimentation Rate (SR) and age-depth model for GC2.

140 cm based on *Gephyrocapsa oceanica* (NN19) and *Emiliania huxleyi* (NN20/21), respectively (Figures 3.1–3.3). This enabled the development of an age model based on depth/age relationship with associated sedimentation rates at the different intervals of study linked to Marine Isotopes Stages, (NN19 = MIS2, NN20 & NN21 = MIS1) relative to the scheme described by Martini (1971), Raffia *et al.* (2006), and Dalibard *et al.* (2014).

The sedimentation rate (SR)/age-depth model investigation of the three GCs shows a similar trend to the MD03–2708C core located in the Ogooué fan, West

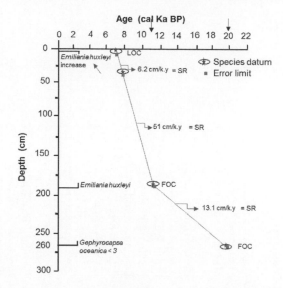

Figure 3.3. Sedimentation Rate (SR) and age-depth model for GC3.

Equatorial Africa (Kim *et al.*, 2010) and the ERCS 17 cores in Niger Delta (Riboulot *et al.*, 2012), where radiocarbon dates have been used to constrain the stratigraphy. The base of the three GCs show similar slumping influx sediments, whereas the middle to uppermost parts of the cored intervals suggest retreating sediments (Figures 3.1–3.3). The relatively constant SR noted in the middle of the GC2 and GC3 cores, and again at the topmost part of the three cores, indicates a similar depositional setting perhaps influenced by local factors such as sea level rise, thus representing a return to reduced palaeodischarge (Barusseau *et al.*, 1995). This study uses derived sedimentation rates (SR) from the confirmed age model to generate data linking the calculation of relative sediment quotients and carried pollen (Figures 4.1–4.3). The results also suggest that it was the similarity of the ages framework between the GCs that assisted with the interpretation of any age/depth related problem.

7.3 RESULTS

This study adopts species grouping into their family/botanical affinities based on the International Palynological Conference, Tokyo (IPC, 2012) (Table 2/Plates 1–3). Due to the high diversification and complex nature of low latitude tropical pollen research in palynology and palaeobotany, most researchers adopt grouping of species into family affinities in order to reflect its vegetational and ecological dynamics rather than assuming a species entity based on their morphology alone (Plates 1–3).

We adopted this method to avoid potential incorrect assignment of identified pollen to an inappropriate ecological or vegetation group due to their extensive diversity and complexity as listed in Table 2. In essence the pollen were first grouped in line with correct morphological parameters to allow assignment with correct family or botanical affinity (e.g. White, 1983; Rull, 1998; Fletcher, 2005; Gosling *et al.*, 2013) before linking their interpretation to develop models concerning prevailing palaeoenvironmental change (Table 2/Plates 1–3).

Table 2. Family and Botanical affinity relating to Morphogenic/Species Nomenclature – example within the same family and vegetation habitat applicable in the GCs study.

Botanical affinity	Pollen/Spore taxon	Family	Vegetation group	Plate no.	Figure no.
Rhizophora spp.	*Zonocostites ramonae*	Rhizophoraceae	Mangrove Swamp	1	16, 18–24
Acrostichum	*Acrostichumsporites* sp.	Pteridaceae	Mangrove Swamp	2	65
cf. *Ipomoea*	*Echiperiporites* sp.	Convolvulaceae	Mangrove Swamp	2	41, 42
Avicennia germinans	*Foveotricolporites crassiexinus*	Avicenniaceae	Mangrove Swamp	2	36, 37
Chamaecrista	?	Fabaceae	Mangrove Swamp	1	38, 39, 40
Cyperus	*Cyperus* sp.	Cyperaceae	Fresh Water Swamp	2	35
Symphonia globulifera	*Pachydermites diederixi*	Clusiaceae	Fresh Water Swamp/ lowland rainforest	1	25–28
Arecaceae	*Psilate/Monocolpites* spp.	Arecaceae	Fresh Water Swamp	1	33
Nymphaea aff. *Lotus*	*Zonosulcites/Psilamonoporites* sp.	Nymphaeaceae	Fresh Water Swamp	2	46
Tabernaemontana	*Psilatricolporites crassus*	Apocynaceae	Lowland Rainforst/ Open forest	1	8, 9
Celtis	*Momipites africanus*	Cannabaceae	Lowland Rainforst/ Open forest	1	12
Anthonothal Berlinial Isoberlinia	*Striatocolpites* sp.	Caesalpinioidiae	Lowland Rainforst	1	30, 31
Amanoa	*Retitricolporites irregularis*	Euphorbiaceae	Lowland Rainforst	1	7
Vernonia	*Spinizonocolpites* sp.	Asteraceae	Lowland Rainforst/ Open forest	1	32
Alchornea aff. *cordifolia*	*Psilatricolporites operculatus*	Euphorbiaceae	Lowland Rainforst	1	14, 15

Fern spores	*Laevigatosporites* sp.	Pteridophyta	Lowland Rainforst/ Open forest	3	66–68
Lycopodium sp.	*Lycopodiumsporites* sp.	Lycopodiaceae	Lowland Rainforst/ Open forest	2	57
Moss spores sp.	*Stereisporites* sp.	Bryophyta	all open habitats	2	60
aff. Hibiscus sp.	*Echiperiporites estalae*	Malvaceae	all open habitats	1	29
Polypodium	*Polypodiaceoisporites* sp. (*Verrucatosporites usmensis*)	Polypodiaceae	Lowland Rainforst/ Open forest	3	69, 70
Borreria (Spermacoce)	*Stephanocolpites* sp.	Rubiaceae	Lowland Rainforst/ Open forest	2	43–45
Fern spores	*Polypodiaceoisporites* sp.	Pteridophyta	Lowland Rainforst/ Open forest	2	62–64
Graminidites sp.	*Monoporites annulatus*	Poaceae	Savannah	1	1–6
Podocarpus	*Podocarpites* sp.	Podocarpaceae	Afromontane	2	47–50

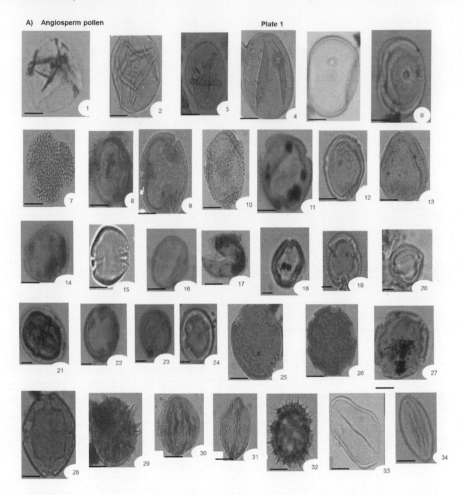

Plate 1. Palynomorphs photomontage, all scale bars are at 20 μm except for Rhizophoraceae which is 10 μm.

Figure	Family name	Morphological name	Figure	Family name	Morphological name
1–6	Poaceae	*Graminidites annulatus*	17	?	*Inaperturotetradites reticulatus*
7	Euphorbiaceae	*Retitricolporites irregularis*	18–24	Rhizophoraceae	*Rhizophora* spp.
8–9	Apocynaceae	*Psilatricolporites crassus*	25–28	Clusiaceae	*Pachydermites diederixi*
10	Euphorbiaceae/ Rubiaceae	*Retibrevitricolporites obodoensis*	29	Malvaceae	*Echiperiporites estelae*
11,13,14	Euphorbiaceae?	*Momipites africanus*	30–31	?Arecaceae/ Anthonotha	*Striatocolpites* spp.
12	Cannabaceae	Celtis	32	Asteraceae	Vernonia-type
15	Euphorbiaceae/ Alchornae	*Psilatricolporites operculatus*	33–34	Arecaceae	*Psilamonocolpites* sp.
16	Rhizophoraceae	*Rhizophora* sp.			

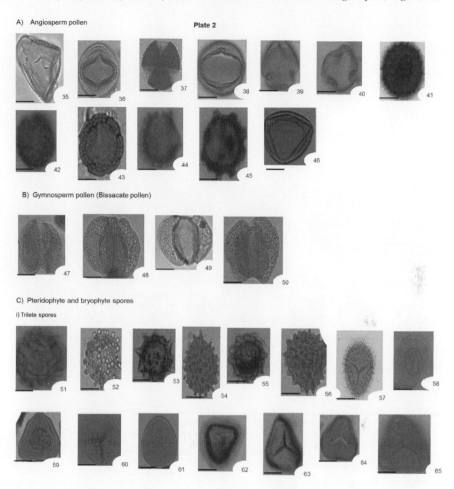

Plate 2. Palynomorphs photomontage, all scale bars are at 20 μm.

Figure	Family name	Morphological name	Figure	Family name	Morphological name
35	Cyperaceae	*Cyperus* sp.	51–56	Asteraceae	*Tubliflorae* sp.
36–37	Avicenniaceae	*Avicennia* sp./ *Foveotricolpites* spp.	52	Arecaceae	*Borassus* type
38–40	Fabaceae	*Chamaecrista* spp.	57	Lycopodiaceae	*Lycopodiumsporites* sp.
41	Convolvulaceae/ cf Ipomoea	*Echiperiporites* sp.	58–59	Polypodiaceae?	*Undulatisporites* sp.
43–45	Rubiaceae/ Spermacoce (Borreria)	*Stephanocolpites* sp.	60	Bryophyta	*Stereisporites* sp.
46	Nymphaceae	*Nyphaea lotus type/ Psilamonoporites* sp.	62–64	Polypodiaceae	*Polypodiaceoisporites* sp.
47–50	Podocarpaceae	*Podocarpidites* sp.	65	Pteridaceae/ Acrostichum	*Acrostichumsporites* sp.
49	Podocarpaceae	*Podocarpidites clarus*			

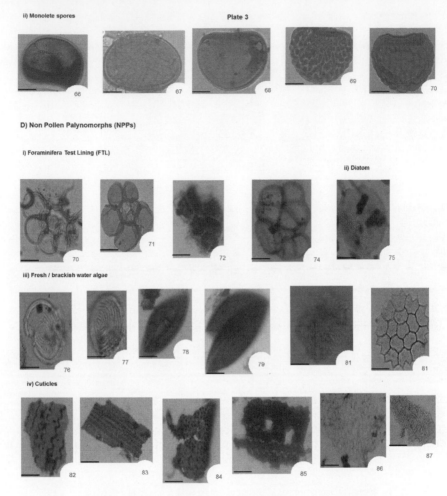

Plate 3. Palynomorphs photomontage, all scale bars are at 20 μm.

Figure	Family name	Morphological name	Figure	Family name	Morphological name
66–68	Pteridophyta	*Laevigatosporites* spp.	78–79	Zygnemataceae	*Ovoidites parvus*
69–70	Polypodiaceae	*Polypodiisporites* sp. *Verrucatosporites usmensis*	80	Brackish water algae	*Botryococcus* sp.
71–72	Foraminifera	Planospiral FTL	81	Fresh water algae	*Pediastrum* sp.
73–74	Foraminifera	Biserial/Triserial FTL	82–85	Graminae	Grass cuticles
75	Diatom	Diatom	86–87	Plantae	Plant cells
76–77	Fresh water algae	*Pseudoschizea* sp.			

Pollen and spores were classified into their relative groups following assessment of their climate, sea level and ecological affinities of the taxa and related parent plants as follows: Mangrove-Coastal Swamp, Freshwater Swamp-Palmae, Lowland Rainforest—Open Forest (semi-arid, sub-humid and humid tropics), Savannah-Afromontane, and Non-pollen palynomorphs (NPPs) (Table 2).

Furthermore, given that sediments may be remobilised and transported from the floodplain and coastal swamp environments and transported to offshore environments, pollen and spores were grouped into two more regional indicators of provenance (Poumot, 1989; Morley, 1995; Rull, 2002): Hinterland (upland) indicators (i.e. fern from lowland rainforest, open forest, savannah-afromontane) and Littoral (lowland) indicators (i.e. mangrove, coastal swamp, freshwater swamp, Palmae).

The hinterland pollen indicators suggest long distance fluvial terrestrial transport of sediment to the marine offshore basin, whereas the littoral pollen indicators suggest short coastal marine sediment transport adjacent to the offshore basin. A full list of species (taxa) indicating botanical affinities and the potential habitat or ecological niche is listed in Table 2 and selectively shown on Plates 1–3. NPPs (non-pollen palynomorphs) were not included among the pollen taxa, with the exception of freshwater algae and cuticle types because of their environmental significance.

7.3.1 Assemblage zonations

Rhizophoraceae as *Zonocostites ramonae* was deliberately excluded from the zones and indicator pollen group Mangrove-Coastal Swamp in the total pollen sum percentage, concentration and flux values to avoid a biased record due to the over representation (Scourse *et al.*, 2005). This approach emphasises the representation of other taxa prevailing climate controls and palaeoenvironmental change.

Four sub-intervals relating to the palynomorph assemblages from the GCs are given from bottom to top depth order to reflect the applied zonation, cluster scheme, and chronology (Figures 4.1–4.3).

7.3.1.1 GC1 Palynomorph intervals

Interval GC1-A1 (275–190 cm)

This interval is characterised by the strong dominance of a hinterland component, namely Lowland Rainforest (Lycopodiaceae, Bryophyta, Pteridaceae, Rubiaceae ~5–40%) as well as an intermediate proportion of Savannah-Afromontane (Poaceae, ~25%), and NPPs (cuticles ~25%).

The pollen concentrations and influx rate from the hinterland component are considered very high, yielding values of 45,000 grains.cm^{-2} and 870 grains.cm^{-2}yr^{-1} respectively. The littoral component representation, especially Mangrove-Coastal and Freshwater Swamp indicators is very low over this interval. Less than ~1% of the Freshwater Swamp indicators group was observed in this interval. The percentage of Rhizophoraceae is low as well (20%).

Freshwater Swamp indicators were also averagely represented with the exception of Cyperaceae were well represented constituting ~ or >20%. The pollen concentrations and influx rates from the littoral vegetation are relatively very low when compared to the hinterland vegetation in this interval, ranging between 120,000 grains.cm^{-2} and 380,000 grains.cm^{-2}yr^{-1} among the indicator groups respectively (Figure 4.1).

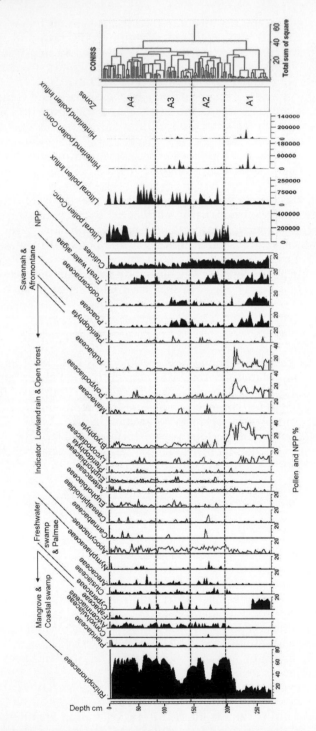

Figure 4.1. Relative Palynomorph Distribution diagram and proposed zonation for GC1. Note that Rhizophoraceae percentages were calculated on a total pollen sum basis, whereas other pollen taxa percentages were calculated on the pollen sum excluding Rhizophoraceae. Footnote on family and species diagram (see Table 2) for more details. This is applicable to both GC2 and GC3 distributions as well.

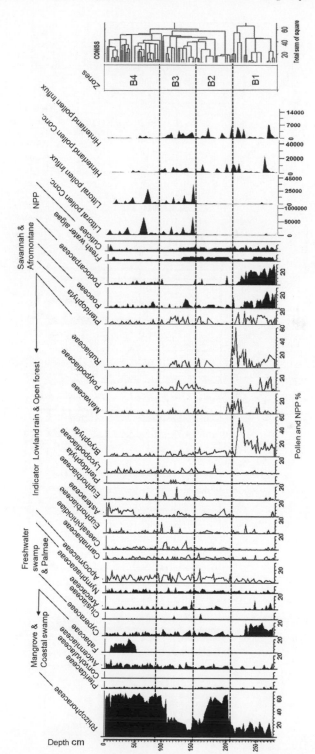

Figure 4.2. Relative Palynomorph Distribution diagram and proposed zonation for GC2.

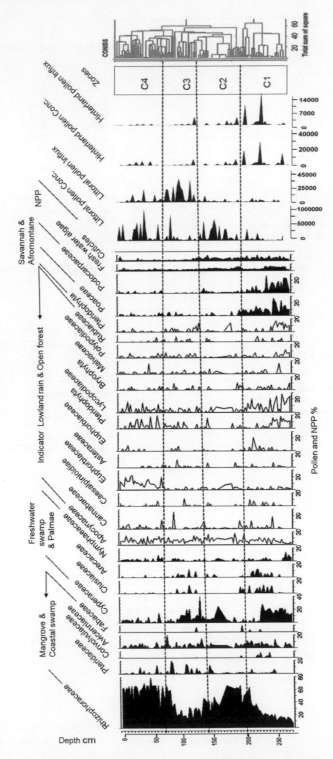

Figure 4.3. Relative Palynomorph Distribution diagram and proposed zonation for GC3.

Interval GC1-A2 (210–145 cm)

This interval is characterised by high values of littoral environmental indicators with a gradual up-hole representation of Mangrove-Coastal Swamp indicators (Rhizophoraceae, Pteridaceae, Convolvulaceae, Avicenniaceae, ~5–75%) dominating, with the close association of low to intermediate representation of Freshwater Swamp indicators (Cyperaeceae, Clusiaceae, Arecaceae and Nymphaeaceae ~3–20%). With the exception of Rhizophoraceae, Avicenniaceae and Cyperaeceae dominate throughout the interval. The relative percentage of Rhizophoraceae increases (75%).

The pollen concentrations and influx rates from the littoral component are high. They consist of 365,000 grains.cm^{-2} and 770,000 grains.cm^{-2}yr^{-1} respectively. The hinterland vegetation over this interval is characterised by the strong reduction of the proportion of Savannah-Afromontane (Poaceae, Podocarpaceae ~2–7%) and NPP (cuticles ~15%). The Lowland Rainforest (Apocynaceae, Asteraceae, Lycopodiaceae, Cannabaceae, Caesalpinioidiae, Euphorbiaceae, Pteridophyta, Euphorbiaceae, Bryophyta, Malvaceae, Polypodiaceae, Rubiaceae ~5–20%) is represented by the appearance of Apocynaceae and Bryophyta.

The pollen concentrations and influx rates from the hinterland vegetation are fairly low. This consists of 93,000 grains.cm^{-2} and 271,000 grains.cm^{2}yr^{-1} within this interval respectively (Figure 4.1).

Interval GC1-A3 (145–85 cm)

Throughout this interval, a different vegetation pattern is observed, which commences with an intermittent increase in the littoral component represented by the gradual expansion of different floral families of the Mangrove-Coastal Swamp vegetation (Rhizophoraceae, Pteridaceae [*Acrostichumsporites*], Convolvulaceae, Avicenniaceae, ~5–65%). This representation is the most prominent and consistent over this interval. The percentage of Rhizophoraceae increases relatively more than other components (~65%).

Freshwater Swamp vegetation representation does not differ much when compared with the preceding Interval GC1-A2 (Cyperaceae, Nymphaceae Clusiaceae, and Arecaceae ~5–15%).

Pollen concentrations and influx rates from the littoral component over this interval are about 300,000 grains.cm^{-2} and 127,000 grains.cm^{-2}yr^{-1}, respectively.

The hinterland component in this interval is characterised by the reduction in the percentage of Lowland Rainforest indicators (Apocynaceae, Asteraceae, Lycopodiaceae, Cannabaceae, Caesalpinioidiae, Euphorbiaceae, Pteridophyta, Euphorbiaceae, Bryophyta, Malvaceae, Polypodiaceae, Rubiaceae ~4–21%). The levels of Savannah-Afromontane vegetation (Poaceae, Podocarpaceae ~3–16%) are similar to the previous intervals. NPPs (cuticles ~8%) decrease, showing a different relationship when compared with the underlying intervals.

The pollen concentrations and influx rates from the hinterland vegetation decrease up-hole. They consist of 61,000 grains.cm^{-2} and 470,000 grains.cm^{-2}yr^{-1}, respectively (Figure 4.1).

Interval GC1-A4 (85–0 cm)

The uppermost interval is characterised by the highest abundance of littoral vegetation, with an abrupt expansion of the different floral families associated with Mangrove-Coastal Swamp vegetation (Rhizophoraceae, Pteridaceae, Avicenniaceae Fabaceae, ~5–75%), except the Convolvulaceae family. The Rhizophoraceae count increases consistently (~80%). The representation of Freshwater Swamp vegetation

(Poaceae and Podocarpaceae ~3–5%). NPPs (cuticles ~4%) decreased in abundance. The pollen concentrations and influx rates from the hinterland vegetation decreased. The values are about 1,000,900 grains.cm^{-2} and 1800 grains.cm^{-2}yr^{-1} respectively (Figure 4.2).

7.3.1.3 GC3 Palynomorph Intervals

Interval GC3-C1 (260–190 cm)
The lowest interval in the GC3 core is also characterised (as seen in both GC1 and GC2) by the prominent presence of the hinterland component consisting of the Lowland Rainforest and Afromontane-Savannah vegetation. The Lowland Rainforest vegetation (Apocynaceae, Asteraceae, Lycopodiaceae, Cannabaceae, Caesalpinioidiae, Euphorbiaceae, Pteridophyta, Euphorbiaceae, Bryophyta, Malvaceae, Polypodiaceae, Rubiaceae ~5–35%) increased more in comparison to other families. Therefore, in this interval the peak of Savannah-Afromontane (Poaceae, Podocarpaceae ~5–25%) increases in proportion, as well as the percentage of Lowland Rainforest vegetation. The percentage of the NPPs (cuticles ~7%) also increases.

The pollen concentrations and influx rates from the hinterland vegetation increase relative to the total pollen count. These consist of 42,000 grains.cm^{-2} and 1800 grains.cm^{-2}yr^{-1}, respectively.

The littoral vegetation begins with a sporadic decline in the Mangrove-Coastal Swamp vegetation indicators (Rhizophoraceae, Pteridaceae, Convolvulaceae, Avicenniaceae ~0–20%). The Rhizophoraceae proportion is low (20%) at this interval.

The percentage of the aquatic taxa, that is, Freshwater Swamp vegetation indicators (Cyperaceae, Nymphaceae, Clusiaceae and Arecaceae ~0–30%) (as hydrophytic adapted plants) increases over this interval proportionately.

The pollen concentrations and influx rates from the littoral vegetation are very low when averages are compared to their equivalent units in the GC1 and GC2 cores. These consist of 27,000 grains.cm^{-2} and 290 grains.cm^{-2}yr^{-1} respectively (Figure 4.3).

Interval GC3-C2 (190–140 cm)
This interval recorded a gradual increase of littoral composition commencing with the Mangrove-Coastal Swamp vegetation indicators (Rhizophoraceae, Pteridaceae, Convolvulaceae, Avicenniaceae ~5–70%). Rhizophoraceae count increases gradually (60–70%) up-hole.

The Freshwater Swamp vegetation indicators (Cyperaceae, Nymphaceae, Clusiaceae and Arecaceae ~3–6%) decline up-hole. The pollen concentrations and influx rates from the littoral vegetation are very high when averages are compared with the other equivalent sections. They consist of 79,000 grains.cm^{-2} and 640 grains.cm^{-2}yr^{-1} values, respectively.

The prominent hinterland component in the Lowland Rainforest vegetation declines, but the pollen associated with these families (Apocynaceae, Asteraceae, Lycopodiaceae, Cannabaceae, Caesalpinioidiae, Euphorbiaceae, Pteridophyta, Euphorbiaceae, Bryophyta, Malvaceae, Polypodiaceae, Rubiaceae ~5–21%) present over this interval increase slightly, when compared with other families in the same group. The Afromontane-Savannah vegetation indicators (Poaceae, Podocarpaceae ~2–5%) also decline along with the NPPs (cuticles ~5%), but Podocarpaceae increases slightly among the two families.

The pollen concentrations and influx rates from the hinterland vegetation decrease. These include 1130 grain.cm^{-2} and 550 grains.cm^{-2}yr^{-1} within this interval respectively (Figure 4.3).

Interval GC3-C3 (140–75 cm)

This interval is characterised by an intermittent decline and increase of littoral contribution beginning with the Mangrove-Coastal Swamp vegetation indicators (Rhizophoraceae, Pteridaceae, Convolvulaceae, Avicenniaceae ~5–40%). Percentage counts of Rhizophoraceae decline slightly as well (40%). The indicators of Freshwater Swamp vegetation (Cyperaceae, Nymphaceae, Clusiaceae and Arecaceae ~5–40%) increase relative to the underlying interval, with particularly Cyperaceae increasing when compared to the previous interval.

The pollen concentrations and influx rates from the littoral component over this interval are fairly high when compared against the average. These include 100,800 grains.cm^{-2} and 26,600 grains.cm^{-2}yr^{-1}, respectively.

The hinterland component consisting of Lowland Rainforest vegetation (Apocynaceae, Asteraceae, Lycopodiaceae, Cannabaceae, Caesalpinioidiae, Euphorbiaceae, Pteridophyta, Euphorbiaceae, Bryophyta, Malvaceae, Polypodiaceae, Rubiaceae ~5–25%) decreased in over this interval except for expansion of Apocynaceae and Euphorbiaceae families. The Afromontane-Savannah vegetation indicators (Poaceae, Podocarpaceae ~2–4%) decreased with alternating higher increase of some hydrophytic adapting pollen (e.g. Asteraceae). In addition, the NPPs (cuticles ~5%) continued to decline as well.

The pollen concentrations and influx rates from the hinterland component decreased and (with a slight increased on the average within the interval), ranging between 760 grains.cm^{-2} and 600 grains.cm^{-2}yr^{-1}, respectively (Figure 4.3).

Interval GC3-C4 (75–0 cm)

The highest interval of the GC3 core begins with the most abundant and noticeable peak of littoral components with an abrupt expansion of the Mangrove-Coastal Swamp vegetation indicators, with the exception of Convolvulaceae and Fabaceae families, which were not recorded (Rhizophoraceae, Pteridaceae, Convolvulaceae, Avicenniaceae ~4–80%). The recovery of Rhizophoraceae increases more when compared to other zones (up to 80%).

The representatives of the Freshwater Swamp vegetation indicators decrease (Cyperaceae, Nymphaceae, Clusiaceae, and Arecaceae ~3–18%), but with all the families being present in this interval.

However, the reappearance of Cyperaceae and Nymphaceae may suggest a different source of vegetation input within this interval. Pollen concentrations and influx rates from the littoral component consist of 900,800 grains.cm^{-2} and 140,800 grains.cm^{-2}yr^{-1}, respectively. Whereas, the hinterland component is characterised by the sharp and consistent decline in the percentages of Lowland Rainforest vegetation (Apocynaceae, Asteraceae, Lycopodiaceae, Cannabaceae, Caesalpinioidiae, Euphorbiaceae, Pteridophyta, Euphorbiaceae, Bryophyta, Malvaceae, Polypodiaceae, Rubiaceae ~2–26%), overall the families increase slightly. It was also noticed that there was a slight peak in Apocynaceae and Euphorbiaceae families when compared with other families in this interval.

Furthermore, there is a consistent decline in the peak of the Savannah-Afromontane vegetation indicators (Poaceae, Podocarpaceae ~2–4%), whereas the percentage of Polypodiaceae decreases slightly when compared to other families. NPPs (cuticles ~4%) consistently decline. The pollen concentrations and influx rates from the hinterland component decrease. They consist of 660 grains.cm^{-2} and 480 grains.cm^{-2}yr^{-1} respectively (Figure 4.3).

7.4 PALYNORMORPH ASSEMBLAGE INTERPRETATION

7.4.1 GCs Palynomorph records

The section below interprets and discusses the palynomorph records from each of the GCs, which have been divided into four intervals, as described in the results section previously. This discussion suggests the relative seasonal changes (climate and sea level) in specific indicator groups linking vegetation patterns in relation to their stratigraphic position within the GCs.

7.4.1.1 GC1 Palynomorph record

NN19 = MIS2 (20–11 ka)
The beginning of the Late Pleistocene is marked by a high representation of taxa (e.g. Podocarpaceae, Poaceae, Lycopodiaceae, Bryophyta, Polypodiaceae, Rubiaceae) that signify cold/dry palaeoenvironment conditions (Figure 5.1). The interval between 275–210 cm in GC1 is accompanied by significant and rapid expansion of Afromontane Forest (Podocarpaceae), Savannah (Poaceae), Lowland Rainforest (Lycopodiaceae, Polypodiaceae, Bryophyta, Rubiaceae), and presence of dry-land derived Charred Grass Cuticle. Cyperaceae and NPPs also increase synchronously with the expansion of this hinterland vegetation. This period indicates a drier/cooler climate with significant presence of open vegetation during the ~20 ka (Dupont *et al.*, 2007; Kim *et al.*, 2010; Marret *et al.*, 2013). Conversely, however, components of Mangrove Swamp vegetation and most of the Freshwater Swamp vegetation, apart from Cyperaceae contribution (thermophilous taxa), decline down-hole sharply during this period. The presence of low mangrove pollen is related to the relative lower sea level (~112–135 m) below the present-day level (Peltier, 1994; Lézine *et al.*, 2005; Punwong *et al.*, 2013).

NN20 = MIS1 (11–8.5 ka)
At the beginning of the Early Holocene, the period is characterised by a gradual rise of Mangrove-Coastal Swamp vegetation (Rhizophoraceae, Avicenniaceae) and a higher representation of pioneer "warm" taxa (Rhizophoraceae), whereas those of Lowland Rain-Open forest, Freshwater Swamp-Palmae (Cyperaceae), Savannah-Afromontane, and Charred Grass Cuticles showed a marked decline, with the exception of Apocynaceae (Figure 5.1). The Step 1 increase (Phase 1) records the onset of the transition between warm/intermediate climate conditions as indicated by a slight decrease of Mangrove vegetation and reappearance of hydrophytic lowland rainforest plants (Dalibard *et al.*, 2014) (Figure 5.1). An episode in which Lowland Rainforest partly replaces less water-dependent Savannah-Afromontane and Freshwater Swamp vegetation was observed at the onset of the warm and relatively wetter period. This period signifies a transition from arid to humid conditions during the early Holocene (Lézine and Vergnaud-Grazzini, 1993; Dupont *et al.*, 1998; Scourse *et al.*, 2005; Marret *et al.*, 2013). The significant reduction in Poaceae, Cyperaceae and Podocarpaceae pollen in the GC1 suggests declines in grassland, herbaceous communities and Afromontane forest (Davis and Brewer, 2009).

NN21 = MIS1 (8.5–6.5 ka)
The beginning of the Mid-Holocene is characterised by a palynomorph record reflecting an expansion in littoral coastal vegetation (Figure 6). This period of rapid increase in Mangrove vegetation trend has been recorded from other African core locations (Dupont and Weinelt, 1996). Step 2 increase (Phase 2) shows a consistent rapid

expansion of Mangrove vegetation and a short-lived expansion of limited "warm" Lowland Rainforest vegetation controlled by the rise of the sea level during the Hypsithermal event (Lézine and Vergnaud-Grazzini, 1993; Scourse *et al.*, 2005).

7.4.1.2 GC2 Palynomorph records

NN19 = MIS 2 (20–11 ka)
Similarly to GC1, the distribution of vegetation patterns and NPPs during the glacial period (Late Pleistocene) in GC2 shows comparable major trends. The beginning of NN19 = MIS2 is also marked by a good representation of the pioneer "cold/dry" taxa, whereas those of "warm/wet" taxa declined similarly to the GC1 profiles (Figure 5.2).

The interval between 266–195 cm is characterised by the expansion in the Afromontane forest (Podocarpaceae), Savannah (Poaceae), Lowland Rainforest (Podocarpaceae, Poaceae, Lycopodiaceae, Bryophyta, Malvaceae, Polypodiaceae, Rubiaceae, Pteridophyta), and Charred Grass Cuticle during the Late Pleistocene. Cyperaceae, Clusiaceae, Arecaceae and NPPs also increase with the expansion of a hinterland vegetation similar to GC1. This period suggests a dry climate, with increased aridity. The increase in abundance of Afromontane and Savannah grassland pollen in the sediments might indicate long distance transport of pollen by enhanced atmospheric circulation (Maley, 1991; Jahns, 1996; Shi and Dupont, 1997). During this time, Mangrove-Coastal Swamp vegetation and most Fresh Water Swamp vegetation (thermophilous taxa), apart from Cyperaceae and Arecaceae, decline rapidly in relation to the postulated lower sea level and correspond well with past records derived from Western Equatorial Africa (e.g. Lézine and Vergnaud-Grazzini, 1993) (Figure 5.2). Dry and cold Afromontane forest development in Central Equatorial Africa has been reported during a similar time period from the Niger Delta and Cameroon (Brenac, 1988; Dupont and Weinelt, 1996).

NN20 = MIS 1 (11–8.5 ka)
The Early Holocene (NN20 = MIS1) is characterised by the gradual rise of Mangrove-Coastal Swamp vegetation (Rhizophoraceae, Avicenniaceae) and an increased establishment of pioneer warm taxa. Step 1 increase records indicate a transition between warm/intermediate climate conditions denoted by the slight decrease of the Mangrove vegetation and reappearance of the hydrophitic Lowland Rainforest plant communities (Dalibard *et al.*, 2014) (Figure 5.2). This period suggests warm climate conditions in relation to sea level rise and local expansion of halophytic vegetation which indicates marsh terrestrialisation and encroachment of brackish environments within lagoonal–marsh settings (Fletcher, 2005, Morley, 1995). During this time-frame, hinterland contributing indicators decreased except for Bryophyta, Malvaceae, Rubiaceae, Apocynaceae (Figure 5.2). It is also a period in which the Lowland Rainforest slightly replaces some of the water-dependent Savannah-Afromontane and Freshwater Swamp vegetation. A significant decline in those vegetation types similar to the GC1 suggests a region under grassland, herbaceous communities and Afromontane forest (Poumot, 1989; Davis and Brewer, 2009) (Figure 5.2).

NN21 = MIS1 (8.5–6.5 ka)
At the onset of the Mid-Holocene, the pollen record in GC2 indicates a two-step relative increase in vegetation changes similar to those recognised in GC1 (Figure 5.2). Step 2 increase in GC2 is slightly different from that observed in GC1, whereby it shows an expansion of Mangrove-Coastal Swamp vegetation (Rhizophoraceae, Avicenniaceae) during sea level rise, coinciding with the Hypsithermal event (Scourse

et al., 2005; Marret *et al.*, 2013; Dalibard *et al.*, 2014). Apocynaceae and Asteraceae slightly increase. This period indicates a wetter climate in relation to the rising sea level flooding the adjacent coastline (Kim *et al.*, 2010).

7.4.1.3 GC3 Palynomorph records

NN19 = MIS 2 (20–11 ka)
As with the records and interpretations for GC1 and GC2, the vegetation and NPPs records during the Late Pleistocene glacial period in the GC3 show a similar increase in dry taxa at the base of the cored sequence. This period (NN19 = MIS2) is represented by a good record of pioneer cold/dry taxa that were abundant during the Late Pleistocene. The warm/wet taxa show a similar decline up-hole as recorded in GC1 and GC2 (Figure 5.3). This period is accompanied by a rapid expansion in the Afromontane forest (Podocarpaceae), Savannah (Poaceae), Lowland Rainforest (Apocynaceae, Bryophyta, Polypodiaceae, Lycopodiaceae, Pteridophyta) and Charred Grass Cuticles during the glacial period.

However, in contrast to GC1 and GC2, Freshwater Swamp indicators and NPPs from this core increase up-hole. Grassland and Open Forest existed along the adjacent continent during that time as a source area for recorded pollen and spores families (Dupont and Agwu, 1991). This period is associated with cold/arid climatic conditions.

Furthermore, as discussed previously regarding the GC1 and GC2, Mangrove Swamp vegetation and most Fresh Water Swamp vegetation (thermophilous taxa) also decline, but with the exception of some of the taxa in GC3 (such as shown on Figure 5.3). The low percentage of mangrove pollen in this period corresponds well with past pollen studies from West Equatorial Africa that identify mangrove expansion, corresponding with exposure of the continental shelf during sea level fall (Dupont *et al.*, 2000; Scourse *et al.*, 2005). Considering the larger recovery of many pollen types belonging to hinterland habitats and the high values of savannah grass pollen found in the modern marine sediments (e.g. Dupont and Agwu, 1991), the percentages of mangrove pollen obtained in the present study during the glacial time are low. This may be due to the arid conditions prevailing during this time.

Similar to the GC1 and GC2, this result from the dry taxa indicates an expansion of Open Forest, Afromontane and Savannah vegetation on the adjacent continent suggesting that a glacial drying/cooling climate prevailed during the last glacial maximum (LGM) in the West Equatorial African regions (e.g. Maley and Brenac, 1998).

NN20 = MIS 1 (11–8.5 ka)
After the glacial period, the Early Holocene (NN20 = MIS 1) interval is marked by the rapid rise of Mangrove-Coastal Swamp vegetation (Rhizophoraceae, Avicenniaceae) and the establishment of warm taxa coinciding with the warm/wet climate conditions during an interglacial time (Rull, 1997). Step 1 increase of the mangrove vegetation indicates an onset of the Early Holocene showing a transition between warm/intermediate climate conditions denoted by a slight decrease of the Mangrove-Coastal Swamp vegetation and a sporadic reappearance of hydrophytic Lowland Rainforest plant components (Dalibard *et al.*, 2014) (Figure 5.3). Similar to the GC1 and GC2, there was also a significant decline in Savannah and Afromontane palynomorph records. These similarities among the three GCs suggest a decline of grassland cover in the region (Poumot, 1989; Rull, 2002; Davis and Brewer, 2009). This arid-humid transition is almost similar to the increase of mangrove pollen in Ogooué Fan core MD03-2708 marking the onset of eustatic sea level rise after the Late Pleistocene (Kim *et al.*, 2010).

NN21 = MIS1 (8.5–6.5 ka)

The Mid-Holocene is characterised again by a two-step increase in vegetation change similar to those documented already for GC1 and GC2 (Figures 5.1 and 5.2). Similar to the GC1 and GC2, the Step 2 increase offers an interpretation for a rapid expansion of the mangrove vegetation during the Hypsithermal event (Scourse *et al.*, 2005; Marret *et al.*, 2013; Dalibard *et al.*, 2014). A similar pattern has been observed from the mangrove pollen off the Niger Delta, Congo and Angola during this time period (Lézine *et al.*, 2005, Scourse *et al.*, 2005; Kim *et al.*, 2005). This expansion in mangrove vegetation indicates a reduced fluvial input and related monsoon circulation punctuated by lower river discharge during the sea level rise (Marret *et al.*, 2013; Punwong *et al.*, 2013).

7.5 DISCUSSION

7.5.1 Preservation and dispersal of palynomorphs

Preservation varies greatly between taxa and is significantly controlled by ecological and climate factors (Fletcher, 2005). These differences are related to the categories of pollination, palaeoecology, and climate factors. In general, a representation of pollen families with increasing distance from the source area is clearly explained (Dafni, 1992), but the actual method of pollen dispersal is controlled by the characteristics of diverse pollen species and the habitat types that control their size, aerodynamic, hydrodynamic, and taphonomic features (Dupont *et al.*, 2007).

Although, preservation of pollen is of great concern for the interpretation of pollen records, in practice, anaerobic sediments have more potential for pollen preservation (Fletcher, 2005). This is because oxygen was insufficient during anoxia, thereby reducing the impact of bacterial degradation on pollen components. On the other hand, the severe influence of bacterial degradation on some pollen types in a sedimentary record due to intensive or rapid decay may result in their absence. If the above statements hold true, it is not surprising to observe from the studied GCs that the Convulaceae and Fabaceae pollen families, which are common littoral pollen in proximal marine records are poorly represented due to biodegradation, whereas Rhizophoraceae, Apocynaceae, Avicenniaceae and Adiantaceae families dominate due to their ability to withstand biodegradation (Fletcher, 2005).

Furthermore, other schools of thought suggest an absence of entomophily (insect pollinated activity) as suggested for the Gulf Coast (e.g. Jardine and Harrington, 2008), as well as the ability of pollen taxa to withstand the rigours of higher sedimentation rates, anemophilous (wind pollinated) effects, restricted high flux of charcoal resulting from wildfires, and burning of the palynomorph reproductory organs during pollen germination (Zobaa *et al.*, 2011). Given these conditions, it could be concluded that Convulaceae and Fabaceae pollen families from the GCs are susceptible to deterioration under such sedimentation rates during lithification or diagenetic processes as well, when compared with other mangrove families (Phuphumirat *et al.*, 2009). Thus, it becomes important to consider these factors in the dispersal and preservation control of palynomorphs in the future for an effective pollen interpretation as discussed above (Figures 5.1–5.3).

7.5.2 Controls of climate drivers on palynomorph records

This section reviews the detailed pollen and vegetation records from the GCs in relation to the impact of climate drivers from the Late Pleistocene to the Mid-Holocene.

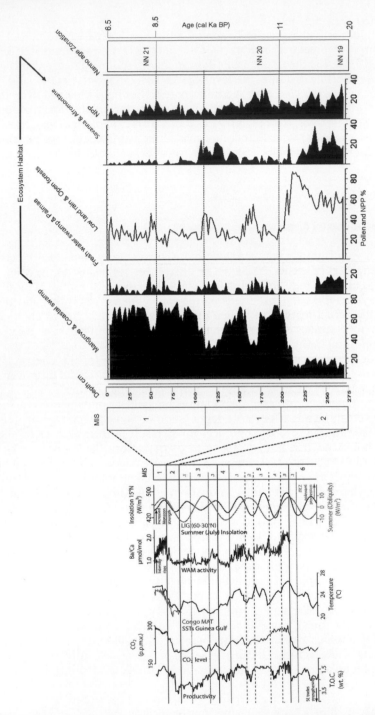

Figure 5.1. Vegetation dynamic diagram of GC1 correlated to MIS 1 & 2 summer LIG 600-300 (Davis and Brewer, 2009); summer insolation curve at 150N (Berger and Loutre, 1991); WAM system curve of Ba/Ca ratio (Weldeab *et al.*, 2007); Gulf of Guinea SSTs (Schneider *et al.*, 1997) Congo Mean Annual Temperature; Vostok CO$_2$ level curve (Petit *et al.*, 1992).

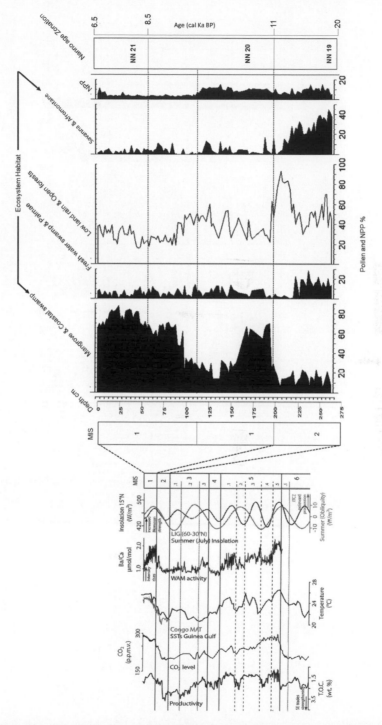

Figure 5.2. Vegetation dynamic diagram of GC2 correlated to MIS1 & 2 summer LIG 600-300 (Davis and Brewer, 2009); summer insolation curve at 150N (Berger and Loutre, 1991); WAM system curve of Ba/Ca ratio (Weldeab *et al.*, 2007); Gulf of Guinea SSTs (Schneider *et al.*, 1997); CO$_2$ curve (Petit *et al.*, 1992).

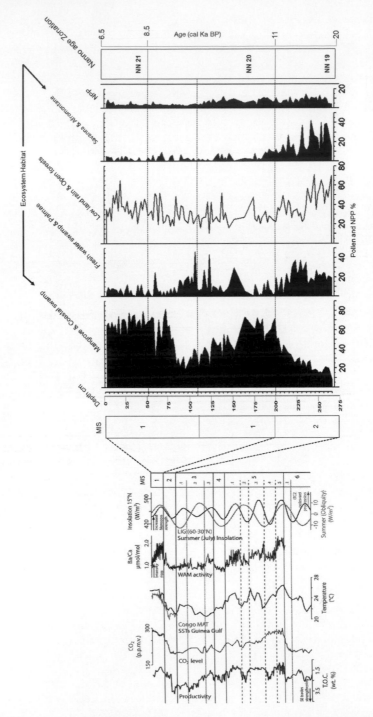

Figure 5.3. Vegetation dynamic diagram of GC3 correlated to MIS 1 & 2 summer LIG 600-300 (Davis and Brewer, 2009); summer insolation curve at 150N (Berger and Loutre, 1991); WAM system curve of Ba/Ca ratio (Weldeab *et al.*, 2007); Gulf of Guinea SSTs (Schneider *et al.*, 1997); CO_2 curve (Petit *et al.*, 1992).

The last glacial (Late Pleistocene) and interglacial periods (Holocene) have been marked with significant shifts between extreme climate conditions (wet and dry) in relation to the dynamics of oceanic and atmospheric circulations (e.g. Hessler *et al.*, 2010). The consequence of this interaction in the North Atlantic causes an increase in the average temperature of the surface waters of the North or South Atlantic Ocean. This effect may coincide with a southward shift of the Intertropical Convergence Zone (ITCZ) (EPICA members, 2006; Dupont *et al.*, 2008; Collins *et al.*, 2010). The temperature relationship between oceans and the atmosphere has influenced vegetation distribution in tropical regions, at least after the post-glacial period (M1 & M2) (e.g. Gasse *et al.*, 2008; Hessler *et al.*, 2010).

Given these relationships (oceanic and atmospheric circulations), the section below discusses how the ITCZ migration and WAM circulations play a dominant role on the distributions of palynomorph records in the tropics of West Africa and their relationship and imprints upon the Niger Delta vegetation dynamics, including their controls on *Rhizophora* and coastal littoral vegetation.

7.5.2.1 Migration of the ITCZ and WAM, and their impacts on the evolution of vegetation in the Niger Delta

Global climate and vegetation configurations are mainly affected by the position and direction of the ITCZ and the strength of the African Monsoon (Figure 2) (Shanahan *et al.*, 2007). Ocean-atmosphere interaction regulates the position of the ITCZ and thus rainfall patterns and river systems ('the hydrological regime') in tropical provinces (e.g. West Africa) (Graham and Barnett, 1987; Leroux, 1993). The current GCs study clearly records a transition from dry to wet climate (East-West across the Niger Delta), which is inferred to coincide with the hypothesis of WAM circulation and ITCZ migration during the Late Pleistocene to Mid-Holocene.

During the last glacial and de-glacial period (20–11 ka), the ITCZ migrated towards the southern hemisphere promoting an arid/dry climate and facilitated the expansion of hinterland vegetation and fluvial transportation of grassland pollen (Hessler *et al.*, 2010) (Figures 4.1, 4.3 and 5.1–5.3). This hinterland expansion demonstrated from the GCs (depths ~272–202 cm) is associated with weak WAM activity presumably causing aridification in the region. In addition to this, the lowland and warm ocean air circulation contributed further to continental aridity influencing dry climates during the Late Pleistocene (Lézine *et al.*, 2005; Kim *et al.*, 2010; Marret *et al.*, 2013). The abundance of charred grass cuticles from the GCs during this period could also be related to an induced wildfire activity, and sequential fluvial discharge through the Benue and Niger Rivers confluence and out across the Niger Delta (Morley and Richards, 1993) (Figures 4.1–4.3).

On the other hand, at the onset of the present interglacial period (11–6.5 ka) the climate patterns become wet/warm due to strengthening of the summer insolation. This causes an increase in WAM activity as the evoked ITCZ migrates northwards towards the equator (Marret *et al.*, 2001). The hinterland vegetation declined during this time (Morley, 1995; Shi and Dupont, 1997; Dalibard *et al.*, 2014) (Figure 6). The expansion of the littoral vegetation during this time, as identified in this study, is similar to the Central and West African records, probably related to rise in sea level in response to melting of ice at polar regions (e.g. Shackleton, 1987; Lézine *et al.*, 2005; Scourse *et al.*, 2005; Marret *et al.*, 2013; Dalibard *et al.*, 2014).

Given this seasonal circulation, transition from dry to wet conditions following the end of the LGM (MIS2), as observed in all three GCs in this study and from other locations (Dupont *et al.*, 2000) in the subequatorial West Africa, is most likely due to the migration of the ITCZ and WAM (e.g. Gasse *et al.*, 2008; Hessler *et al.*, 2010; Weijers *et al.*, 2007; Marret *et al.*, 2013). The vegetation records recognised through pollen and NPPs of the GCs could serve to evaluate the relationship between monsoonal and ITCZ dynamics. The coherence from the interpretation of the palynomorph records of GCs exclusively substantiates previous evidence, and has also contributed to the "recent dialogue on the strength and importance of the tropical monsoon" in controlling the regional climate of West Africa (Guilderson *et al.*, 1994; Blunier *et al.*, 1998).

7.5.2.2 Controls on Rhizophora and coastal littoral vegetation of the Niger Delta

Generally, mangrove pollen grains are considered to be good stratigraphic markers of sea level changes within the tropical regions of the world. Pollen produced by mangrove species is often present in shallow subtidal and intertidal settings (e.g. Scourse *et al.*, 2005).

Further to the previous discussions (ref. Results), this section highlights the importance of mangrove data and what has actually stimulated or controlled the strongly comparative increases across the three GCs. The similar expansion in the mangrove pollen in the GCs (GCs depth of ~202–0 cm) is due to vegetation change as a result of sea level rise during the Early to Mid-Holocene, and erosion and reworking of old mangrove sediments that were deposited on the sea floor of the Niger Delta (Morley, 1995) (Figure 6). The observed increase in mangrove pollen recovery is not exclusively due to forest expansion, but due to transportation and re-deposition of previous mangrove pollen sediments onto the continental shelf during sea level rise (Van Campo and Bengo, 2004; Scourse *et al.*, 2005; Kim *et al.*, 2005; Dupont *et al.*, 2007) (Figure 6) as in other regions like Australia, mangrove expansion has been termed the "big swamp phase" as a result of sea level rise (Schmidt, 2008). Moreover, this episode could suggest a regional or local control (e.g. uplift) of enhanced wet climate patterns influencing the Niger Delta sediments coinciding with the strengthening of the summer insolation and intense WAM system (Morley, 1995; Lézine and Denfle, 1997).

However, it is possible that the slight decline in the *Rhizophora* pollen data (GCs depth of ~175–75 cm) after the Early Holocene, close to the Mid-Holocene could be due to the influence of intermittent or gradual sea level rise transgressing the shoreline, or perhaps, other local factors that our data cannot sufficiently explain in this study (Ellison, 2000) (Figure 6). It appears that the presumed decline in mangrove vegetation after the Early Holocene is not simply evidence of a change in the vegetation but rather a relative reduction in the graphical peak resulting from lack of higher recovery of *Rhizophora* pollen within those intervals (Figure 6). This is because previous controlled analysis resulted in the shortage of sample material bulk rock volume to be analysed within specific intervals (10 intervals). Inadequate sample was left for further analysis in order to compensate for the shortfall of the mangrove data (Figure 6, especially GCs 2 & 3 *Rhizophora* pollen plots).

7.5.2.3 Impact of climate change on the littoral mangrove ecosystems of the Niger Delta

This section addresses the impact of a warm climate on sea level by using the expansion of the coastal mangrove vegetation as a proxy. Case studies on mangrove ecosystems along the West African coastline from 1997 to 2015 can be used as a modern analogy to the Quaternary dataset in this section in relation to the fourth objective of this study. The present deforestation, reclamation, and water logging within the

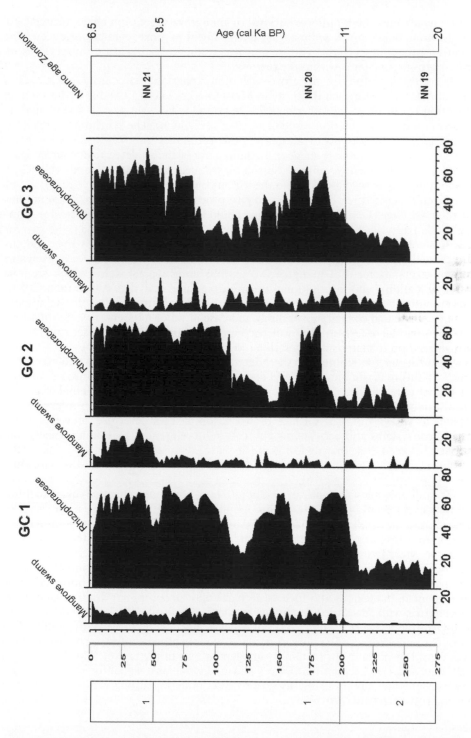

Figure 6. Diagram demonstrating similar phase of Mangrove fluctuations (% calculated on a total sum excluding Rhizophoraceae) and with Rhizophoraceae percentages for the three GCs.

region has affected the long-term survival of mangrove vegetation characterised by a very dynamic but complex sedimentary environment in most regions worldwide. The concern for the future is how to utilise previous studies to assist future conservation of the mangrove ecosystems within the region.

In context, this study on mangrove vegetation expansion derived from the GCs (GCs depth of ~202-0 cm) through the Early to Mid-Holocene (11–6.5 ka) due to the sea level rise during the warm to hypsithermal climate period indicates a past period of marine incursion on the coastline (Punwong *et al.*, 2013) (Figure 6). The synchronous relationship between the expansions of mangrove pollen from the three GCs across the East-West Niger Delta presumably suggests a uniform rise of the sea level, regional responses to environmental influences, similar controls on sediment accumulation, and evidence of dominant mangrove vegetation within the region (Figure 6) (Adojoh *et al.*, 2016).

Given that the "present is the key to past processes" (Lyell, 1830, p. 33), it implies that the past Niger Delta palaeo-shoreline could be interpreted using indicator mangrove pollen as well as to forecast future sea level scenarios. To this effect, the current study has confirmed the dynamic response of coastal vegetation to sea level and climate conditions in the Niger Delta after 11 ka, and reinforces the need for coastal and sustainable management of mangrove vegetation along coastal zones, which are now centres of national and international research planning activities (Ogadinma, 2013). Consequently, increasing awareness of global warming relative to its potential impact on sea level rise and coastal morphology, increase in water temperature, and quantity of the continental or terrestrial runoff becomes a significant question meriting further study in relation to mangrove vegetation (Nicholls and Cazenave, 2010).

As a further contribution to the Nigerian economy, it is imperative to assess the impact of future climate and sea level change on the Niger Delta because of the current global consequences resulting in the risk of flooding, environmental degradation, drought, and for the future sustainability of the region based on interpolating evidence as shown from the GCs at 11–6.5 ka on past sea level rise. However, since climate, geosystems and ecosystems control the development of human society, the feedback between these driving forces is critical to creating predictive scenarios for the development of the socio-economic system of the region (IPCC, 2013; Massuanganhe *et al.*, 2016; Adojoh *et al.*, 2016).

Overall, this study provides an insight into how a changing climate is likely to influence the development of ecosystems through past study on sea level rise at 11.7–6.5 ka resolved from the depth within the GCs ~ 202-0 cm (Figure 6). With the on-going aggressive flooding resulting from the impact of global warming affecting the ecosystems within the Niger Delta region (e.g. influence on the Eleme and Bonny refineries due to their locations/position), these findings hint at the need to consolidate and expand the initiatives to protect this valuable coastal and vegetational ecosystem, observed to be ever present in the region, and focused in particular on combining traditional conservation efforts with adaptation to climate impact in the future (Nicholls and Cazenave, 2010; Ogadinma, 2013). Conservation of ecosystems (mangrove vegetation) should be of utmost importance to the present decision-making process affecting marine coastal systems and their ability to sustain future generations in the region. This becomes an important necessity because mangroves play a key role as a signal for recognising sea level transgression, ability to swiftly absorb CO_2, buffering strong winds, and offering coastal protection (BBC, 2016).

7.6 CONCLUSIONS

The main hypotheses tested were the interactions between the position of the ITCZ and the strength of the WAM on explaining fluctuations in the palynomorph records from selected gravity cores located in the Niger Delta. Our data support the hypothesis that the of ITCZ and WAM have played a leading role on the control on vegetation evolution and coastal shift of the Niger Delta, similar to other regions in West Africa (e.g. Sanaga, Congo, Senegal and Ogooué deep sea deltas). Palynomorph distribution in marine delta sediment indicates a good correlation in latitudinal occurrence of sporomorphs (pollen and spores) between terrestrial source and the marine sink, providing the basis for reconstructing different biomes over the selected time frame.

Based on the interpretation of pollen assemblages within the sediment cores it can now be concluded that the hinterland versus littoral pollen plots from the three GCs records have prompted an understanding of an ecosystem response (vegetation dynamics) to the various controls (dry versus wet climate, sea level, and edaphic factors) due to the alteration of the climate drivers (ITCZ and WAM). It is suggested that the interpreted expansion of the hinterland vegetation within this region is coincident with arid/cool climate, low sea level, and weak monsoonal conditions in relation to the southerly position of the ITCZ during the 20–11 ka (NN19 = MIS2) period, whereas the expansion of littoral vegetation is linked to sea level rise during humid/wetter climate conditions in relation to northern migration of the ITCZ at 11–6.5 ka (NN20 and NN21 = MIS1).

Given these conditions, it is important to emphasise that the similar relationship obtained from the GCs palaeovegetation (palynomorph) records provides strong evidence of the timing of arid and humid transitions during the Late Pleistocene–Mid Holocene, respectively. This causative relationship using palaeovegetation proxies, family affinity, and climate drivers has been made in the Niger Delta region for the first time. It can now be confirmed and concluded that the controls on palaeovegetation dynamics of the Niger Delta region is linked to the timing of the establishment of the WAM in relation to the migration of the ITCZ (Figure 2).

ACKNOWLEDGEMENTS

This work is part of a PhD thesis sponsored by the Petroleum Technology Development Fund of Nigeria (PTDF). Without their funding, this research would have been difficult for me (OA) to undertake in an overseas environment.

Many thanks to both Lydie Dupont of Bremen University, Germany who reviewed and significantly contributed to family affinities of the vegetation groupings and to Carlos Jaramillo of the Smithsonian Research Group, USA, for his assistance in terms of confirmation of some of the pollen taxa. The SPDC Geological Services team (Port Harcourt) are thanked for the supply of the three gravity cores and other materials to enable the research program to happen.

Finally, we extend our appreciations to Dr. Greg Botha and one anonymous reviewer for their cogent comments.

REFERENCES

Adojoh, O., Fabienne, M., Robert, D., Peter, O., 2016, Relative impact of sea level change and sediment supply on shallow offshore Niger delta margins from palynodebris and lithofacies data. *Quaternary International*, **404**, p. 178.

Adojoh, O. and Osterloff, P., 2010, Phytoecological reconstructions of Eocene-Pliocene sediments, Niger Delta. 3rd NAPE/AAPG Regional Deepwater African Offshore Conference, Abuja. Abstract A 6.

Barusseau, J.P., Bâ-diar, A.M., Descamps, C., Diop, H.S., Giresse, P. and Saos J.l., 1995, Coastal evolution in Senegal and Mauritania at 103, 102 and 101-year scales: Natural and human records. *Quaternary International*, **29**, pp. 61–73.

BBC, 2016, British Broadcasting Corporation News on mangrove vegetation in Sri Lanka (http://www.bbc.com/news/science-environment).

Berger, A. and Loutre, M.F., 1991, Insolation values for the climate of the last 10 million years. *Quaternary Science Reviews*, **10**, pp. 297–319.

Blunier, T., Chappellaz, J., Schwander, J., Dallenbach, A., Stauffer, B., Stocker, S.T., Raynaud, D., Jouzel, J., Claussen, H.B., Hammer, C.U. and Johnsen, S.J., 1998, Asynchrony of Antarctic and Greenland climate change during the last glacial period. *Nature*, **394**, pp. 739–743.

Bouimetarhan, I., Dupont, L., Kuhlmann, H., Pätzold, J., Prange, M., Schefuß, E. and Zonneveld, K., 2015, Northern Hemisphere control of de-glacial vegetation changes in the Rufiji uplands, Tanzania. *Climate of the Past*, **11**, pp. 751–764.

Brenac, P., 1988, Evolution de la végétation et du climat dans l'Ouest-Cameroun entre 25 000 et 11 000 ans BP. Actes X Symposium Association des Palynologues de langue française, Inst. Fr. Pondicherry, *Science Technology*, **25**, pp. 91–103.

Clemens, S.C., Prell, W.L. and Sun, Y., 2010, Orbital-scale timing and mechanisms driving Late Pleistocene Indo-Asian summer monsoons: Reinterpreting cave speleothem $\delta^{18}O$. *Paleoceanography*, **25**, 4207. doi: 10.1029/2010PA001926.

Collins, J.A., Schefusz, E., Heslop, D., Mulitza, S., Prange, M., Zabel, M., Tjallingii, R., Dokken, T.M., Huang, E., Mackensen, A., Schulz, M., Tian, J., Zarriess, M. and Wefer, G., 2010, Interhemispheric symmetry of the tropical African rainbelt over the past 23,000 years. *Nature Geosciences*, **4**, pp. 42–45.

Dafni, A., 1992, *Pollination ecology: A practical approach*, (Oxford: Oxford University Press).

Dalibard, M., Popescua, S., Maley, J., François Baudin, F., Melinte-Dobrinescue, M., Pittetf, B., Marsset, T., Dennieloug, B., Droz, L. and Succ, J., 2014, High-resolution vegetation history of West Africa during the last 145 ka. *Geobios*, **46**, pp. 183–198.

Davis, B.S. and Brewer, S., 2009, Orbital forcing and role of the latitudinal insolation/temperature gradient. *Climate Dynamics*, **32**, pp. 143–165.

Doust, H.E. and Omatsola, E.M., 1990, Niger Delta. In *Divergent/Passive Basins*, edited by Edwards, J.D. and Santagrossi, P.A., *American Association of Petroleum Geologists Bulletin*, **45**, pp. 201–238.

Dupont, L.M. and Agwu, C.O., 1991, Environmental control of pollen grain distribution patterns in the Gulf of Guinea and offshore NW-Africa. *International Journal of Earth Sciences*, **80**, pp. 567–589.

Dupont, L.M. and Agwu, C.O., 1992, Latitudinal shifts of forest and savanna in N.W. Africa during the Brunhes chron: Further marine palynological results from site M16415 (9°N, 19°W). *Vegetation History and Archaeobotany*, **1**, pp. 163–175.

Dupont, L.M., Behling, H. and Kim, J.H., 2008, Thirty thousand years of vegetation development and climate change in Angola (Ocean Drilling Program Site 1078). *Climate of the Past*, **4**, pp. 107–124.

Dupont, L.M., Behling, H., Jahns, S., Marret, F. and Kim, J.H., 2007, Variability in glacial and Holocene marine pollen records offshore from west southern Africa. *Vegetation History and Archaeobotany*, **16**, pp. 87–100.

Dupont, L., 2011, Orbital scale vegetation change in Africa. *Quaternary Science Reviews*, **30**, pp. 3589–3602.

Dupont, L.M., Jahns, S., Marret, F. and Ning, S., 2000, Vegetation change in equatorial West Africa: Time-slices for the last 150 ka. *Palaeogeography, Palaeoclimatology, Palaeoecology*, **155**, pp. 95–122.

Dupont, L.M., Marret, F. and Winn, K., 1998, Land-sea correlation by means of terrestrial and marine palynomorphs from the equatorial East Atlantic: phasing of

SE trade winds and the oceanic productivity. *Palaeogeography, Palaeoclimatology, Palaeoecology*, **142**, pp. 5–84.

Dupont, L.M. and Weinelt, M., 1996, Vegetation history of the savanna corridor between the Guinean and the Congolian rain forest during the last 150,000 years. *Vegetation History and Archaeobotany*, **5**, pp. 273–292.

Ellison, J.C., 2000, How South Pacific mangroves may respond to predicted climate change and sea level rise. In *Climate change in the South Pacific: Impacts and responses in Australia, New Zealand, and Small Islands States*, edited by Gillespie, A. and Burns, W., (Dordrecht, Netherlands: Kluwer Academic Publishers), pp. 289–301.

EPICA Community Members, 2006, Eight glacial cycles from an Antarctic ice core. *Nature*, **429**, pp. 623–628.

Fletcher, W.J., 2005, Holocene Landscape History of Southern Portugal. Doctoral thesis. Cambridge University Press, pp. 159–172.

Gasse, F., Chalié, F., Vincens, A., Williams, M.A.J. and Williamson, D., 2008, Climatic pattern in equatorial and southern Africa from 30,000 to 10,000 years ago reconstructed from terrestrial and near-shore proxy data. *Quaternary Science Reviews*, **27**, pp. 2316–2340.

Gehrels, W.R., 1999, Middle and Late Holocene sea-level changes in Eastern Maine reconstructed from foraminiferal saltmarsh stratigraphy and AMS 14C dates on basal peat. *Quaternary Research*, **52**, pp. 350–359.

Gosling, W., Miller, C. and Livingstone. D.A., 2013, Atlas of the tropical West African pollen flora. *Review of Palaeobotany and Palynology*, **199**, pp. 1–135.

Graham, N.E., and Barnett, T.P., 1987, Sea surface temperature, surface wind divergence and convection over tropical oceans. *Science*, **238**, pp. 657–659.

Griffiths, J.F., 1972, *Climates of Africa*, (Amsterdam: Elsevier), **9**, pp. 195.

Grimm, E.C., 1987, CONNISS, a FORTRAN 77 programme for stratigraphically constrained cluster analysis by the method of incremental sum of squares. *Computers and Geoscience*, **13**, pp. 13–35.

Guilderson, T.P., Fairbanks, R.G. and Rubenstone, J.L., 1994, Tropical temperature variations since 20,000 years ago: Modulating interhemispheric climate change, *Science*, **263**, pp. 663–665.

Hessler, I., Dupont, L., Bonnefille, R., Behling, H., González, C., Helmens, K.F., Hooghiemstra, H., Lebamba, J., Ledru, M.P., Lézine, A.M., Maley, J., Marret, F. and Vincens, A., 2010, Millennial-scale changes in vegetation records from tropical Africa and South America during the last glacial. *Quaternary Science Reviews*, **29**, pp. 2882–2899.

IPC, 2012, International Palynological Congress and Palaeobotany Conference, Tokyo, Japan. *Conference Proceedings*, Special Issue, pp. 58.

IPCC, 2013, Climate Change 2013: The physical science basis. Contribution of working group I to the fourth assessment report of the Intergovernmental Panel on Climate Change, (Cambridge: Cambridge University Press).

Jahns, S., 1996, Vegetation history and climate changes in West Equatorial Africa during the Late Pleistocene and Holocene based on a marine pollen diagram from the Congo fan. *Vegetation History and Archaeobotany*, **5**, pp. 207–213.

Jardine, J. and Harrington, G., 2008, A diverse floral assemblage from the late Palaeocene of Mississippi. *Palynology*, **32**, pp. 261–262.

Kim, S.Y., Scourse, J., Marret, F. and Lim, D.I., 2010, A 26,000-year integrated record of marine and terrestrial environmental change off Gabon, west equatorial Africa. *Palaeogeography, Palaeoclimatology, Palaeoecology*, **297**, pp. 428–438.

Kim, J.H., Dupont, L., Behling, H. and Versteegh, G.J.M., 2005, Impacts of rapid sea-level rise on mangrove deposit erosion: application of taraxerol and *Rhizophora* records. *Quaternary Science*, **20**, pp. 221–225.

Kulke, H., 1995, Nigeria, In *Regional petroleum geology of the World: African, America, Australia and Antarctica*, edited by Kulke, H., (Berlin: Gebruder Bornbraeger), **11**, pp. 143–172.

Leroux, M., 1993, The mobile polar high: A new concept explaining present mechanisms of meridional airmass and energy exchanges and global propagation of palaeoclimatic changes. *Global and Planetary Change*, **7**, pp. 69–93.

Lézine, A.M., Cazet, J.P. and Duplessy, J.C., 2005, West African monsoon variability during the last deglaciation and the Holocene: Evidence from fresh water algae, pollen and isotope data from core KW31, Gulf of Guinea. *Palaeogeography, Palaeoclimatology, Palaeoecology*, **219**, pp. 225–237.

Lézine, A.M. and Denèfle, M., 1997, Enhanced anticyclonic circulation in the eastern North Atlantic during cold intervals of the last deglaciation inferred from deep-sea pollen records. *Geology*, **25**, pp. 119–122.

Lézine A.M. and Vergnaud-Grazzini, G., 1993, Evidence of forest extension in West Africa since 22,000 BP: A pollen record from the eastern tropical Atlantic. *Quaternary Science Reviews*, **12**, pp. 203–210.

Lyell, C., 1830, *The principles of geology*, (London: Murray), vol. 2, Chapter 2.

Maley, J., 1991, The African rain forest vegetation and palaeoenvironments during the late Quaternary. *Climate Change*, **19**, pp. 79–98.

Maley, J. and Brenac, P., 1998, Vegetation dynamics, palaeoenvironments and climatic changes in the forests of western Cameroon during the last 28,000 years B.P. *Review of Palaeobotany and Palynology*, **99**, pp. 157–187.

Marret, F., Kim, S.Y. and Scourse, J., 2013, A 30,000 ka yr record of land–ocean interaction in the eastern Gulf of Guinea. *Quaternary Research*, **80**, pp. 1–8.

Marret, F., Scourse, J., Versteegh, G., Jansen, J., Fred, S. and Ralph, R., 2001, Integrated marine and terrestrial evidence for abrupt Congo River palaeodischarge fluctuations during the last deglaciation. *Journal of Quaternary Science*, **16**, pp. 761–766.

Martini, E., 1971, Standard Tertiary and Quaternary calcareous nannoplankton zonation. In *Proc. 2nd Int. Conf. Planktonic Microfossils*, Roma, (Rome: Ed. Tecnosci), edited by Farinacci, A., **2**, pp. 739–785.

Massuanganhe, E.A., Westerberg, L.O., Risberg, J., Preusser, F., Bjursäter, S. and Achimo, M., 2016, Geomorphology and landscape evolution of Save River delta, Mozambique. pp. 1–21.

Morley, R.J., 2000, *Origin and evolution of tropical rainforest*, (Chichester: Wiley), pp. 288–299.

Morley, R.J., 1995, Biostratigraphic characterisation of systems tracts in Tertiary sedimentary basins. *Proceedings of International Symposium on Sequence Stratigraphic in SE Asia*, pp. 50–71.

Morley, R.J. and Richards, K., 1993, Gramineae cuticles: A key indicator of Late Cenozoic climatic change in the Niger Delta. *Review of Palaeobotany and Palynology*, **77**, pp. 119–127.

Nicholls, R.J. and Cazenave, A., 2010, Sea-level rise and its impact on coastal zones. *Science*, **6**, pp. 329–628.

Ogadinma, A., 2013, To assess the hydrological impact of climate change on Niger Delta, Nigeria. In *2nd International Conference on Earth Science and Climate Change*, Las Vegas, USA, **4**, pp. 69.

Peltier, W.R., 1994, Ice age paleotopography. *Science*, **265**, pp. 195–201.

Petit, J.R., Jouzel, J., Raynaud, D., Barkov, N.I., Barnola, J.M., Basile, I., Bender, M., Chappellaz, J., Davisk, M., Delaygue, G., Delmotte, M., Kotlyakov, V.M., Legrand, M., Lipenkov, V.Y., Lorius, C., Pepin, L., Ritz, C., Saltzmank, E. and Stievenard,

M., 1992, Climate and atmospheric history of the past 420,000 years from the Vostok ice core, Antartica. *Nature*, **399**, pp. 429–436.

Phuphumirat, W., Dallas C.M. and Choathip, P., 2009, Pollen deterioration in a tropical surface soil and its impact on forensic palynology. *The Open Forensic Science Journal*, **2**, pp. 34–40.

Poumot, C., 1989, Palynological evidence for eustatic events in the tropical Neogene. *Centres for Research Exploration Production Elf Aquitaine*, **13**, pp. 437–453.

Punwong, P., Marchant, R. and Selby, K., 2012, Holocene mangrove dynamics and environmental change in the Rufiji Delta, Tanzania. *Vegetation History and Archaeobotany*, **12**, pp. 1–16.

Punwong, P., Marchant, R. and Selby, K.A., 2013, Holocene mangrove dynamics from Unguja Ukuu, Zanzibar. *Quaternary International*, **298**, pp. 4–19.

Raffia, I., Backman, J., Fornaciari, E., Pälike, H., Rio, D., Lourens, L. and Hilgen, F.J., 2006, A review of calcareous nannofossil astrobiochronology encompassing the past 25 million years. *Quaternary Science Reviews*, **25**, pp. 3113–3137.

Reijers, T.J.A., 2011, Stratigraphy and sedimentology of the Niger Delta. *Geologos*, **17**, pp. 133–162.

Riboulot, V., Cattaneo, A., Berné, S., Schneider, R., Voisset, M., Imbert, P. and Grimaud, P., 2012, Geometry and chronology of Late Quaternary depositional sequences in the Eastern Niger Submarine Delta. *Marine Geology*, **319–322**, pp. 1–20.

Rull, V., 1997, Quaternary palaeoecology and ecological theory. *Bolten de la Sociedad Venezolana de Geologos*, **46**, pp. 16–26.

Rull, V., 1998, Middle Eocene Mangroves and Vegetation Changes in the Maracaibo Basin, Venezuela. *PALAIOS*, **13**, pp. 287–296.

Rull, V., 2002, High impact palynology in petroleum geology. Applications from Venezuela (northern South America). *American Association of Petroleum Geologists Bulletin*, **86**, pp. 279.

Schmidt, D., P. 2008, A palynological and stratigraphic analysis of mangrove sediments at Punta Galeta, Panama. (Ph.D., University of California, Berkeley) **3331783**, pp. 2–29.

Schneider, R.R., Price, B., Muller, P.J., Kroon, D. and Alexander, I., 1997, Monsoon related variations in Zaire (Congo). Sediment load and influence of fluvial silicate supply on marine productivity in the east equatorial Atlantic during the last 200,000 years. *Paleoceanography*, **12**, pp. 463–481.

Scourse, J., Marret, F., Versteegh, G.J.M., Jansen, J.H.F., Schefuss, E. and van der Plicht, J., 2005, High resolution last deglaciation record from the Congo fan reveals significance of mangrove pollen and biomarkers as indicators of shelf transgression. *Quaternary Research*, **64**, pp. 57–69.

Shackelton, N.J., 1987, Oxygen isotopes, ice volume and sea level. *Quaternary Science Reviews*, **6**, pp. 183–190.

Shanahan, T.M., McKay, Nicholas, P.H., Konrad A., Overpeck, J.T., Otto, B., Bette, H., Clifford, W., King, J., Scholz, C.A. and Peck, J., 2015, The time-transgressive termination of the African Humid Period. *Nature Geoscience*, **8**, pp. 140–144.

Shanahan, T.M., Overpeck, J.T., Sharp, W.E., Scholz, C.A. and Arko, J.A., 2007, Simulating the response of a closed-basin lake to recent climate changes intropical West Africa (Lake Bosumtwi, Ghana). *Hydrological Processes*, **21**, pp. 1678–1691.

Shell, 2011, Location map of Niger Delta and core positions (GC cores).

Shi, N. and Dupont, L., 1997, Vegetation and climatic history of southwest Africa: a marine palynological record of the last 300,000 years. *Vegetation History Archaeobotany*, **6**, pp. 117–131.

Sowunmi, M.A., 1981, Aspects of Late Quaternary vegetational changes in West Africa. *Journal of Biogeography*, **8**, pp. 457–474.

Tjallingii, R.M., Claussen, J.B.W., Stuut, J., Fohlmeister, A., Jahn, T., Bickert, F. and Röhl, U., 2008, Coherent high and low-latitude control of the northwest African hydrological balance. *Nature Geoscience*, **1**, pp. 670–675.

Van Campo, E. and Bengo, M.D., 2004, Mangrove palynology in recent marine sediments off Cameroon. *Marine Geology*, **208**, pp. 315–330.

Weijers, W.J., Schefub, E., Schouten, S. and Sinninghe Damsté, J.S., 2007, Coupled thermal and hydrological evolution of tropical Africa over the last deglaciation. *Science*, **315**, pp. 1701–1704.

Weldeab, S., Lea, D.W., Schneider, R.R. and Andersen, N., 2007, 155,000 years of West African Monsoon and Ocean Thermal Evolution. *Science*, **316**, pp. 1303–1307.

White, F., 1983, The vegetation of Africa: A descriptive memoir to accompany the UNESCO/AETFAT/UNSO vegetation map of Africa. *Natural Resources Research*, **20**, pp. 1–356.

Zobaa, M.K., Zavada, M.S., Whitelaw, M.J., Shunk, A.J. and Oboh-Ikuenobe, F.E., 2011, Palynology and palynofacies analyses of the Gray Fossil Site, eastern Tennessee: Their role in understanding the basin-fill history. *Palaeogeography, Palaeoclimatology, Palaeoecology*, **308**, pp. 433–444.

CHAPTER 8

Palaeoenvironments and Palaeoclimates during the Upper-Pleistocene and Holocene in the western Lake Kivu region

Chantal Kabonyi Nzabandora
Faculty of Sciences, University of Bukavu, D.R. Congo

Emile Roche
Palaeontology, University of Liège, Belgium

Mike I. Akaegbobi†
Department of Geology, University of Ibadan, Nigeria

ABSTRACT: Palaeoenvironmental reconstruction of the inter-tropical regions of Africa have been carried out based on palynological studies. However, there are little or no palaeoenvironmental information on the areas west of Lake Kivu in spite of several approaches over the last twenty years. Sedimentary deposit of the Cishaka sequence was sampled on the foot of Mount Kahuzi at 2260 m a.s.l. in the Kahuzi-Biéga National Park (KBNP) located on the western Rift Valley flank of the Lake Kivu basin (D.R. Congo) and subjected to palynological investigation. The palynological data allowed the reconstruction of 35,000 years of regional palaeoenvironmental history. About 60 organic sediment samples derived every 10 cm from a 6 m depth core in the Cishaka sequence were palynologically analysed after extraction of fossils by acetolysis procedure. The sediment samples are mainly dark grey coloured clays. The base of the core is poor in pollen grains and has been radiocarbon dated up to 38,800 ± 4100 BP.

Based on different forest taxa, nine phases or zones of forest evolution were established. The dynamics of the main components of the forest taxa within the zones was used to subdivide the first zone into three subzones and the zone 7 into two subzones. During the Upper Pleistocene ("Kalambo Interstadial", 32,000–26,000 years BP), a very heterogeneous forest occupies the area under a moderately hot and humid climate. During the regressive phase of the "Hypothermal" of Mount Kenya (25,000–15,000 years BP, LGM), vegetation cover shows an important extension of herbaceous open environment that witnesses a distinct dryness.

Subsequently, the restoration of an Afromontane forest with an Afro-subalpine connotation, point towards cold and humid climatic conditions which was temporarily interrupted by a renewed extension of more open environments attesting a dry throbbing that can be linked to the Younger Dryas (YD). During the African Humid Period (AHP, 10,000–7000 years BP) precipitation increased. Lake Kivu showed an important rise in sea level, whereas the depressions in altitude were transformed into shallow lakes. The subsequent progression of Afromontane rainforest that probably followed has not been archived because of an insufficient sedimentological record in the Cishaka sequence. Around 6500 years BP, the first swampy deposits of the Holocene reveal the existence of an Afromontane mixed rainforest with a mesophilic character. Around 4000 years BP, an important shift in natural climatic conditions (not anthropogenically driven!) caused open environments, highland grasslands at high altitude, and savannas on lower slopes. This indicates a hydrous deficit that can be put in relation with the African global period of aridity to this time ("First Millennium Crisis"). After an instable climatic situation till 2000 years BP, the ecosystem gained stability, like today's conditions which allowed forests during the first millennium AD to recolonize the area.

8.1 INTRODUCTION

Among the inter-tropical regions of Africa where palynological studies had been carried out with regards to the reconstruction of depositional palaeoenvironments, little has been published on the densely vegetated lap-sided zones of Lake Kivu. The main objective of the present work is to reconstruct the forest environment of the Kivu region during the last 35,000 years. This study focuses on the "high altitude" sector of the Kahuzi-Biéga National Park (KBNP). This is because the climatic factors are favourable to the growth of peaty deposits generated within a natural environment of a highland ombro-mesophile forest. This research attempts to reconstruct the evolution of the surrounding forest environment and to identify different evolutionary phases in a chronological setting. It also aims to distinguish between the climatic and anthropogenic influences that shaped the Afro-highland landscape from the Quaternary up to the present. This environment presents its own specificity because, contrary to other oriental-centre sites as in Rwanda, it is located in a permanent watered zone with regular humidity. This specificity requires the application of specific methodological approaches to refine the interpretation of the results. The palynological survey was carried out at a peat site called Cishaka in the Mount Kahuzi sector of the KBNP.

8.1.1 Geographical, geomorphological and environmental framework

The Kivu region is located to the east of the Democratic Republic of the Congo and it is dominated in its extreme eastern end by the western ridge of the East African Rift System (EARS) (Runge, 1997, 2001). The Mitumba mountain chain culminates at 3310 m a.s.l. at the Mount Kahuzi. The Kahuzi-Biéga National Park, covering an area of 6000 km², was created in 1970 to protect gorillas of the eastern plains (*Gorilla beringei graueri*) and their habitat. It includes two main summits, namely Kahuzi (3310 m a.s.l.) and the Biéga (2790 m a.s.l.). This National Park is located in the province of South Kivu, in the southern part of the mountain chain northwest of the town of Bukavu (2°30'50" S, 28°50'37" E), the capital site of the province (Figure 1).

8.1.2 Location of study site

The study site can be accessed through forest tracks at the start of the Bukavu-Kisangani road. After leaving the park gate at Tshivanga (2100 m a.s.l.), the road crosses a ridge at 2398 m a.s.l. (Figure 1c). Between the latter and the foot of the Mount Kahuzi, a vast plateau extends over a notched faulted plain which is covered by large swamps. All visited sites are covered by paludicole vegetation with dominance of *Cyperus denudatus* and *Cyperus latifolius*. The marshy area of Cishaka is located at 2260 m a.s.l. within a surrounding ombro-mesophile forest located between 2000 and 2400 m a.s.l. Rainfall on this western backbone of Kivu is important. Going up the slopes of Mount Kahuzi one encounters a bamboo zone (*Arundinaria alpina*) that is included in the upper floor of the forest. From 2800 m up to the summit at 3300 m a.s.l., it changes into a more open vegetation cover with herbaceous trees and an Afro-subalpine to Afro-alpine vegetation type (Páramo), where precipitation is less, but frequent mists are persistent, which maintains a quasi-permanent atmospheric moisture. The Cishaka swamp is the second largest of the prospected study area, after the Musisi-Ngushu swamp. It can be reached by road (4 km behind the park gate), then on foot through the forest at a distance of 2.5 km to the north (2°17'34" S, 28°44'24" E; 2260 m a.s.l.). A sedimentary sequence of 6 m was manually cored. Composed of a succession of peats, clayey peats, peaty clays, and organic clays more or less compact (Figure 2). The base of the sequence was dated using ^{14}C age dating up to 38,800 ± 4100 years BP (GrN – 32518). Its upper part (–35 cm) is composed mainly of less consistent fibrous peat and decomposed organic materials.

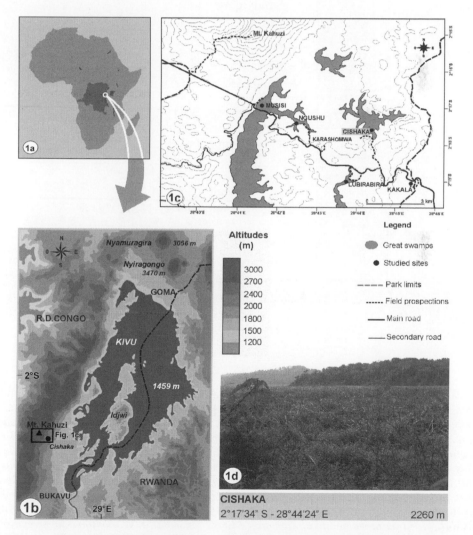

Figure 1. Location of the study area showing the sample sites and drill points.
1a. Location in Africa, **1b.** Regional location, **1c.** Field location (map), **1d.** Landscape view of the swamp.

CISHAKA- 2260 m / 2°17'34" S - 28°44'24" E

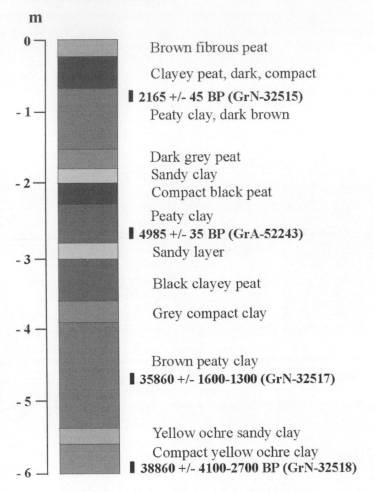

Figure 2. Stratigraphic sketch of the Cishaka sequence.

The vegetation within this swamp is mainly associated with Cyperaceae and composed of *Alchemilla ellenbeckii, Asplenium sandersonii, Begonia meyeri-Johannis, Brillantaesia cicatricosa, Brillantaesia patula, Christella gueintziana, Crassocephalum vittelinum, Gallium chloronoianthum, Impatiens stulmannii, Mikaniopsis tedlei, Osmunda regalis, Peperomia fernandopoiana, Pilea johnstoni, Pilea rivularis, Phyllanthus odontadenius, Rubus steudneri, Rhynchostigma racemosa, Spermacoce latifolius,* and *Triumfetta cordifolia.* The adjoining areas around the swamp site is dominated by a mixed Afromontane forest with *Syzygium rowlandi* (dominant), *Canthium guenzii, Carapa grandiflora, Galiniera coffeoides, Harungana madagascariensis, Hypericum revolutum, Macaranga neomildbraediana, Maesa lanceolata, Neoboutonia macrocalyx, Rapanea melanophloeios,* and *Symphonia globulifera.* Sporadically isolated plants of *Podocarpus milanjianus, Olea capensis,* and *Sinarundinaria arundinacea* occur.

Table 1. Radiocarbon dates of Cishaka peat samples.

Location	Sample Depth (m)	Lab.-No.	Age (Years BP)
Cishaka (Kivu-DRC)	−0.80 m	GrN–32515	2165 ± 45
	−2.80 m	GrA–52243	4985 ± 35
	−4.80 m	GrN–32517	35,860 ± 1600
	−6.00 m	GrN–32518	38,860 ± 4100

8.2 SAMPLE MATERIALS AND METHODS

Utilizing a "Russian" sampling corer, sediment samples were collected at different levels where organic matter was favourable for dating. In total, a sedimentary sequence of 6 m was penetrated.

A description of the marsh vegetation and their surroundings was made on the basis of own botanical inventories. The samples derived at various depth were labelled and described in the laboratory, and one to two cm³ of sediment were collected every 10 cm for analysis. A set of four samples was prepared for ^{14}C age dating (Table 1).

8.2.1 Preparation of samples

The extraction of the fossil organic material was carried out by standard process of acetolysis technique developed by Erdtman (1960). This involves, firstly, the dehydration of the sample by cold acetic acid followed by treatment with a mixture of nine volumes of acetic anhydride and one volume concentration of sulfuric acid. The whole set up is heated in a sand bath for 15 minutes and then diluted with distilled water. The mixture is filtered through a sieve with mesh size of 200 µ and, thereafter, a filtration with a filter of 12 µ is carried out for samples containing mineral particles (e.g. silica). This process is preceded by HF (40%) treatment.

8.2.2 References for interpretation

Reconstruction of the regional palaeoenvironmental evolution is based on references of phytosociological studies specifically concerning the study area. Despite the quality of recent publications (Fischer, 1996; Habiyaremye, 1997), reference is mainly made to older detailed studies to develop and refine the interpretation (Lebrun, 1935, 1936; Lebrun and Gilbert, 1954). Indeed, although recognized in 1980 as a UNESCO world heritage owing to its biodiversity, the KBNP, because of human depredations, is now seriously degraded and therefore devaluated (Kabobyi Nzabandora *et al.,* 2011). It is, thus, on the basis of all aforesaid works that the results of Cishaka were interpreted.

8.3 RESULTS

The palynological analysis from the sequence of Cishaka covers only five metres of depth because the last metre of the profile was very poor in pollen grains. The evolution of the different forest taxa reveals nine phases of Afromontane forest development (Figure 3). Zone 1 was subdivided into three subzones, and zone 7 into two subzones as there was little botanical evolution in the main zones.

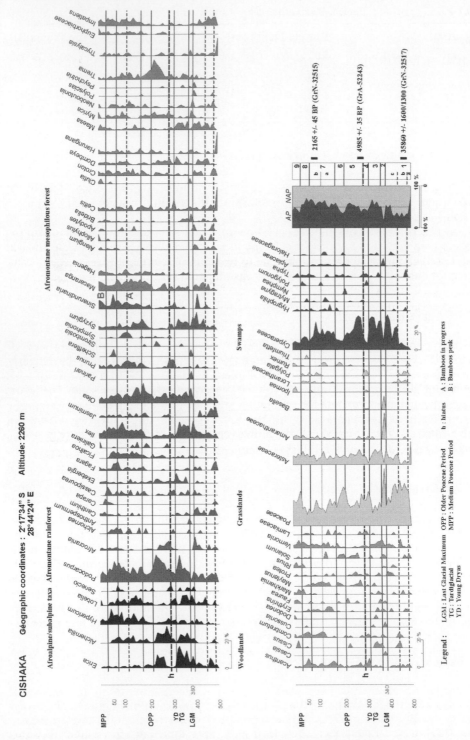

Figure 3. Pollen diagram of the Cishaka sequence.

8.3.1 Zone 1: From 500 cm to 390 cm

Zone 1 is assigned to a ^{14}C age date range from $35,860 \pm 1600$ BP (GrN-32517) to –480 cm and it reflects the evolution of a forest setting in expansion that is achieved in three stages.

8.3.1.1 Subzone 1a: From 500 cm to 490 cm

This first subzone rather reveals an association of woody elements that are not disparate, where it is recorded as dominant taxa *Hagenia* (20%), *Celtis* (20%), *Harungana* (20%), and *Trycalysia* (20%) in an open environmental context where Asteraceae (20%) and Poaceae (28.96%) successively dominate and dense environmental component has only few meaningful representatives such as *Podocarpus, Fagara, Ilex, Jasminum, Olea*, and *Syzygium*.

8.3.1.2 Subzone 1b: From 490 cm to 450 cm

In this subzone, the forest environment shows a tendency to structuring. There is an increase of *Erica* (5.13%), *Podocarpus* (4.95%), *Canthium* (1.75%), *Jasminum* (3.58%) and *Ilex* (2.98%); while the mesophile component of the environment is still dominant, except for *Macaranga, Allophylus, Apodytes, Maesa*, and *Vernonia*, which are the main elements, whereas *Hagenia* strongly retreats. The Poaceae (31.51%) are still in progress indicating an open environment, whereas Asteraceae (2.20%) regress and Cyperaceae (7.43%) are stagnant at a very low level for a swampy environment.

8.3.1.3 Subzone 1c: From 450 cm to 390 cm

Here a heterogeneous forest is already established with a tendency to tighten up itself. Poaceae pollen indicates a significant shrinking and the ombrophile forests start to extend. *Podocarpus* remained relatively stable and the progress is especially noticed among *Afrocrania* (6.32%), *Canthium* (2.79%), *Cassipourea* (1.20%), *Ilex* (10.84%), *Syzygium* (12.19%), and *Sinarundinaria* (12.29%). In these mesophilic assemblage an increase was observed for *Macaranga* (4.20%), *Croton* (1.74%), *Dombeya* (2.94%), and *Maesa* (4.18%). It should also be mentioned that *Polyscias* progress. The strong progression of the Cyperaceae (51.06%) is followed by of an increase of *Erica, Hypericum* and *Lobelia*. By the end of the zone 1, the arboreal pollens (AP), which previously dominated are now replaced by non arboreal pollens (NAP).

8.3.2 Zone 2: From 390 cm to 380 cm

This depth range represents a short period with a strong increase ("explosion") of Poaceae (48.42%), Amaranthaceae (22.63%), and *Basella* (12.63%). The Afro-subalpine component with *Alchemilla* (3.61%), *Erica* (6.32%), and *Lobelia* (1.58%) occurs simultaneously whereas a general decrease of Cyperaceae (4.50%) is striking. The majority of the woody elements drastically decrease. However, maintenance or even increase of some taxa has been noted for, e.g. *Bridelia* (3.61%), *Cassia* (1.05%), *Clutia* (0.35%), *Cussonia* (0.53%), *Harungana* (1.20%), *Myrica* (2.11%), *Olea* (7.89%), *Parinari* (2.11%), *Schefflera* (1.05%), and *Vernonia* (3.65%). In zone 2, the NAPs clearly dominate the APs very distinctly.

8.3.3 Zone 3: From 380 cm to 310 cm

This zone is characterized by a new conquest of the largely open grass area stemming from zone 2 by a set of taxa of the ombrophile dense forest and of the Afro-subalpine taxa of which some show a significant increase, e.g. *Podocarpus* (17.16%), *Ilex* (9.47%), *Syzygium* (6.11%), *Alchemilla* (8.20%), *Erica* (13.92%), and *Lobelia* (2.70%). In that evolution, there is a modest increase of *Cassipourea* (1.26%), *Ekebergia* (4.55%), and *Sinarundinaria* (4.55%) as well as the ongoing distribution of *Olea* (4.14%) within the sample. The mesophilic component develops very little, as well as the woody assemblage indicating a more open landscape. However, it can be stated that a slow progression of *Macaranga* (2.37%), *Dombeya* (4.55%), *Maesa* (4.17%), and *Trema* (3.30%); and the shrinking of *Vernonia* (1.37%) is taking place. Asteraceae (9. 66%) and Cyperaceae (49.16%) pollens are in progress, while Poaceae (7.30%) decline.

8.3.4 Zone 4: From 310 cm to 300 cm

At the beginning of zone 4, a decline of Cyperaceae (4.35%) is recognizable, similar to that described within pollen zone 2 (no proportionately increase of Poaceae (27.27%)). On the contrary, there was a noticeable increase of the Asteraceae (11.07%), including of *Senecio*. With the decrease of the swamp area, a total shrinking of *Erica* (0.34%) is recognizable, but it is less pronounced for *Alchemilla* (15.71%) and *Lobelia* (1.75%). Except *Afrocrania* (3.57%), *Prunus* (4.80%), and *Myrica* (1.57%) that are in progress and *Olea* (3.76%) as well as *Macaranga* (1.37%) that are steady, all the forest genres such as *Podocarpus, Cassipourea, Ekebergia, Syzygium, and Sinarundinaria* are decreasing, especially *Ilex* (1.43%). At the end of this pollen zone, Cyperaceae (47.35%) resume increasing seriously in unison with *Alchemilla* (15.71%). Only *Podocarpus* (7.87%) undergoes a renewal/rejuvenation within the former forest environment.

8.3.5 Zone 5: From 300 cm to 240 cm

This section of the core was ^{14}C dated to 4985 ± 35 Years BP (GrA-52243) at −290 cm. It attests more to a forest recolonization of the Afro-highland type which is rather superior and underlines an important increase of *Podocarpus* (17.65%) with a discrete development of *Ficalhoa* (1.50%); *Olea* (4.81%) remaining steady. The mesophilic components, *Hagenia* (1.46%) is also present and *Myrica* (1.53%) maintains itself. The only representative taxon of this component is *Trema* (16.78%) whose extension has been particularly marked during the mid-Holocene. The previously open environment retreats as the Poaceae (15.24%) retrogress. The measured decline on Cyperaceae (17.58%) is followed by the striking presence of *Alchemilla* (8.23%) and a progression of *Erica* (7.19%).

8.3.6 Zone 6: From 240 cm to 200 cm

This zone corresponds to a significant decline of the tree pollens, whereas the rate at which Poaceae (30.51%) is represented is still particularly high. *Podocarpus* (4.29%) undergoes a retreat that is not negligible in an ombrophile setting, while *Olea* (12.78%) and *Syzygium* (6.02%) continue to progress within the landscape. The only important movement in mesophilic fraction is the shrinking of *Trema* (1.47%) succeeded by a progression of *Macaranga* (4.80%) and *Hagenia* (2.02%). Cyperaceae (11.72%) undergo a sudden fall in a swamp environment, whereas *Alchemilla* (1.65%), Erica (1.21%), and *Lobelia* (0.44%) also decrease.

8.3.7 Zone 7: From 200 cm to 90 cm

Zone 7 is indicative of the marked presence of an association of Afro-highland taxa of the dense forest integrating some affirmed Afro-subalpine elements and the mesophilic species largely committed to the whole system.

8.3.7.1 Subzone 7a: From 200 cm to 140 cm

This zone corresponds to the progress of *Podocarpus* (8.86%) associated to *Ficalhoa* (2.76%), that is consistent as well as *Ilex* (5.06%), *Prunus* (3.10%), *Symphonia* (1.43%) and *Syzygium* (4.43%) while *Olea* (3.23%) shows a recession. *Sinarundinaria* whose two peaks (3.49% and 3.50% respectively) are perceptible and follows the progression of *Podocarpus,* but it is noted that its fluctuations appear in alternation with the latter. An integral part of the dense ombrophile rainforest is noticed by the presence of *Alchornea, Canthium, Cassipourea, Ekebergia, Fagara, Galiniera, Jasminum.* As for *Macaranga* (6.45%) and *Neoboutonia* (2.15%), they come out of the lot of a mesophilic assembly without any particularity evidence. Cyperaceae (28.86%) are in expansion, similar to *Alchemilla* (6.45%), *Hypericum* (6.91%), especially *Lobelia* (10.05%) and contrarily to Poaceae (11.52%) that are slowly receding, whereas the percentage of Asteraceae (3.80%) pollens remain stable.

8.3.7.2 Subzone 7b: From 140 cm to 80 cm

At the more recent water level of the swamp, Cyperaceae (42.49%) continue to progress as do *Podocarpus* (11.47%) and *Sinarundinaria* (11.98%). They show another distribution peak/maximum (2165 ± 45 years BP (GrN - 32515) at 80 cm depth level, while within the forest environment little variation seems to have occurred. Apart from *Ilex* (8.89%) in progress, the stability of *Fagara* (2.79%), *Hagenia* (3.37%) and *Myrica* (1.67%) and that of *Alchornea* and *Macaranga* is also recorded, along with the shrinking of *Prunus* and *Syzygium.*

8.3.8 Zone 8: From 80 cm to 50 cm

In zone 8, the ombrophile rainforest components represent a certain instability, while the mesophile hardly changed. Some of the forest taxa occur less such as *Podocarpus* (6.94%), *Fagara* (0.28%), *Ilex* (5.33%), and *Prunus* (0.44%). This decline is particularly significant for *Olea* (1.98%) and *Sinarundinaria* (5.58%). Other elements like *Alchornea, Carapa,* and *Cassipourea* are progressing. This seemingly apparent instability does not influence any opening of the environment, with the Poaceae (12.54%) remaining limited to relatively low rates. The swampy zone seems to shrink, because of the weak percentage of Cyperaceae (11.74%) whereas *Hypericum* (10.16%) expand.

8.3.9 Zone 9: From 50 cm to 32.5 cm

Zone 9 evokes an evolution of the environment in which the NAPs take to step over the APs. Poaceae (27.71%) shows an important extension, especially at the beginning of this zone; while the majority of the woody species declines, except for some heliophile genres like *Macaranga, Hagenia, Vernonia,* and *Sinarundinaria* (5.66%) that show a little peak while the swamp did not evolve. The expansion of Cyperaceae (6.29%) remains at very low rates.

8.4 INTERPRETATION OF RESULTS

8.4.1 Palaeoenvironmental evolution of the western Rift Valley

The palynological study covers the core at −500 cm, a level that is radiocarbon dated around ca. 36,000 years BP (35,860 years BP at 4.80 m, Table 1). Pollen zone 1 reveals the evolution of an initially open forest environment that is shrinking during the Holocene. However, this dynamic process to form today's vegetation cover took some time to structure itself. The first two pollen subzones cannot be interpreted as a typical Afro-highlander environment as it is currently known.

At the beginning of a heterogeneous concealing grouping of the taxa that are spread in the open surroundings, like *Hagenia, Harungana, Fagara,* and even *Trycalysia,* a pollen assembly is represented where mesophile species, notably *Macaranga, Alangium, Allophylus, Apodytes, Dombeya, Maesa* and *Vernonia* form a vegetation mosaic of the occupation of an increased land; whereas ombrophile rainforest components are still scattered.

Within subzone 1c, a complex of mixed and structured ombro-mesophile forest appears in which the ombrophile fraction dominates the mesophile one. During this period, herbaceous species are ousted and receded. *Podocarpus, Afrocrania, Canthium, Ilex, Syzygium* and *Sinarundinaria* are in meaningful progression and attest the constitution of a medium vegetation belt (Lebrun, 1935, 1936) of the Afromontane forest in which *Macaranga* and *Maesa* progress moderately. The development of the swamp or shallow lake is emphasized by an important extension of Cyperaceae attended by a progression of *Syzygium guineense/rowlandi* species that form the border vegetation of marshes and swamps, while the occurrence of *Ilex* would result from an increase in humidity forming a favourable environment for them, and, so far, leading to the development of the heliophile genus. The pollen peak of Bamboos could indicate a temporary decrease of the temperature and the humidity during this period.

In pollen zone 2, the maximal expansion of herbaceous plants, mainly Poaceae, but also of Amaranthaceae (*Sericostachys*) and *Basella,* heliophile genus, invade swamp borders and glades which would indicate a local opening of the landscape. This also evidenced by the receding of most of the woody elements. However, it can be stated that an increasing of taxa such as *Erica, Alchemilla, Lobelia, Senecio,* and *Shefflera* was related to a slightly colder climate. *Olea* and *Myrica* usually cover open spaces showing water deficit, while *Cussonia, Harungana* and *Vernonia* benefit of the free spaces created by these particular conditions. Therefore, this would indicate an environment shaped by a climatic deterioration with both, decrease of temperature, and reduced humidity. The low percentage of Cyperaceae indicates that occasional dryness affected the swampy marshes seriously. This cold and dry period could correspond to a sudden and sharp phase of the climatic deterioration during and around the Last Glacial Maximum (LGM).

Pollen zone 3 can be distinguished from the foregoing by the growth of the ombrophile and Afro-subalpine elements; mesophile elements are continuously regressing, as well as the Poaceae. The taxa of zone 3, conjointly progressing with that of *Podocarpus,* reflect vegetational aspects developing under cold and humid climatic conditions. There are some enrolled elements associated with a certain ambient humidity. They spread in altitude like *Ekebergia, Ilex, Syzygium* (sp. *parvifolium*) and *Trema;* these last three being rather heliophiles. The progression of *Maesa,* associated with one of the *Erica,* and the presence of *Hypericum* could also be the expression of a cold and humid climate (Lebrun and Gilbert, 1954). The little significant increasing of *Sinarundinaria* suggests that the cold climate would have been a little too low to encourage its development. The expansion of the swamp/shallow lake could result from these particular climatic conditions underlined by a progression of *Alchemilla*, *Lobelia* and of *Syzygium* component of the peri-swampy vegetation.

The pollen assemblages of zone 4 denote again a period of climatic deterioration. It is shown that the NAPs newly dominate the APs and the only woody genres that are not receding are those that have the potential to tolerate not only lower temperatures such as between 8° to 10°C (up to 3000 m a.s.l.), but also a serious reduction of rainfall (–200 mm/year). These taxa have also the capability to spread into the more open surroundings. If *Olea* resists, maintaining its previous percentages, the progression of *Afrocrania, Cassipourea, Prunus, Myrica, Senecio* and Asteraceae in general can still be noticed. The drought influences the extent of the swamp, as suggested by the weak percentages of Cyperaceae and *Alchemilla*, near of those found in zone 2. The comparison between the zones 2 and 4 is not so meaningful, however, for Poaceae whose progression is modest. Poaceae's presence has the tendency to prove that the climatic conditions are less cold. The short "phase" of a forest deterioration shown in zone 4 is typical for transitional climates from cold-humid to cold-dry, creating an opening of vegetation and savanna-like landscapes. It is hypothesized that this period could correspond to the Younger Dryas (YD) well known from high latitudes.

Around 5000 years BP (pollen zone 5), a severe forest progression under the influence of a cold mountain climate with elevated atmospheric humidity (mist/fog) occurred. The pollen storage reveals an assemblage of the upper horizon type of the highland forest in positive evolution where *Podocarpus* and *Trema* arc the most dynamic essences. *Macaranga*, usually present besides *Podocarpus* in the regressive areas, is replaced here by *Trema*, a very dynamic and fast growing colonizer of glades and borders (in the superior belt of the mountain forest, Lebrun, 1935, 1936; Lebrun and Gilbert, 1954), in progressive series of a moderately open environment as well as the rates of Poaceae give evidence for it. Despite the ambient humidity, a relative decrease of precipitation could explain the receding of the swamp, invaded by *Alchemilla*. Other expansive areas would be occupied by *Erica*. The expansion of *Podocarpus* and *Erica* suggests some rigorous climatic conditions, especially resulting from lower mean annual temperatures (Ntaganda, 1991).

The spatial expansion of Poaceae and the predominance of the NAPs on the APs characterize pollen zone 6. Jointly, a pronounced receding of forest taxa is noticed, except for *Olea, Prunus, Syzygium, Macaranga* and *Hagenia* that are in progress. The increase of *Macaranga* and *Hagenia* indicates a regression of dense forest, contrary to the one found into pollen zone 5, where *Trema* was particularly dynamic. The regressive evolution could be assigned to possible climatic disruptions such as a cool fluctuation associated with a "hydrological crisis", as the increase and possible spatial extension of *Olea* suggests. As for *Prunus* and *Syzygium* (probably *S. parvifolium*), they are species susceptible to adjust themselves, like adopting a shrivelled shape. They can adapt to modified and "difficult" climatic conditions (Combe, 1977). The considerable regression of the swamp, in all its components, would be consecutive to this period of a drier climate that could be assimilated to the Older Poaceae Period (OPP), an episode suggested by Runge (2001a) intervening around 4000 years BP.

Pollen zone 7 indicates an expansion of the dense Afro-highland forests evidenced by Afro-subalpines and mesophile taxa elements largely associated with it. These assemblages depend upon a cool climate with regular but not heavy rains. The coexistence of *Podocarpus* and *Ficalhoa*, with the steady installation of *Macaranga* as well as the progression of *Sinarundinaria* can be noticed.

In subzone 7a, the setting is dominated by *Podocarpus*, associated with *Ficalhoa, Ilex, Prunus, Symphonia, Syzygium, Macaranga*, and *Neoboutonia*. It points towards general climate conditions becoming more humid again (Marchant and Taylor, 1998; Habiyaremye, 1997). The associated taxa of this whole subzone 7a are essentially ombrophiles such as *Alchornea, Canthium, Cassipourea, Ekebergia, Fagara, Galiniera,* and *Jasminum.* The renewal of humidity could explain the receding of *Olea.* Hence, the mesophile elements seem to be less important. During this period, two peaks of *Sinarundinaria,* probably consecutive of a light climatic variability, are observed. Retreat of Poaceae suggests the narrowing of the open areas, while the progression of Cyperaceae points out a swamp (higher water table) in expansion, subsequently invaded by *Alchemilla* and *Lobelia* in humid and open spaces against *Hypericum* and *Maytenus* in the drier and mainly wooded or forested zones.

In subzone 7b a sensible progression of *Podocarpus* and *Sinarundinaria* appears, which reaches a new peak of extension, 2100–2200 years BP, more important than the precedents. The simultaneous increasing of the *Ficalhoa* emphasizes its coexistence with the two aforementioned genres in a mixed grouping developing within the dense massive forest. In the evolution of this subzone, an extension of some genera such as *Alchornea, Fagara, Ilex, Macaranga* is recorded, whereas others genera like *Prunus, Symphonia, Syzygium* and *Olea* decrease. This environmental aspect suggests a slight rising of the temperature and humidity. During this period, a momentarily open environment would justify the peak of Poaceae that is observed as well as the progression of heliophile genera such as *Ilex, Myrica, Polyscias,* and *Trema.* The posterior increasing of *Schefflera* could also be related to the upcoming distribution of Bamboos. The evolution of the swamp follows the same schema, that is, a receding after the extension of Cyperaceae. If the stability of *Alchemilla* and *Lobelia* is recorded, *Hypericum* is in distinct progress, which could mean that a recession of species would have occurred after the phase of expansion resulting from the temporary opening of the spaces and the expansion of Bamboos fostering erosion.

Pollen zone 8 constitutes the introduction of the environmental setting of the last two millennia. Among the progression of the ombrophile taxa, a light increase of *Carapa, Cassipourea* and *Ilex* is noticed. These three taxa—of which the last is heliophile and the two others usually dominated within the dense Afromontane forest—benefitted from a largely open space that attests the progression of the Poaceae. A slight progression is also recorded for *Hagenia, Croton, Vernonia,* and *Impatiens,* all four colonizers of glades. After its previous regression, the swamp stabilizes itself; *Hypericum* is always largely dominant there. Globally, to this level, the NAPs and the APs are more or less equivalent. One can suggest that the forest regression observed in Zone 9 records the upset of a climatic instability. This s confirmed in the pollen spectrum of Zone 9 by a major expansion of the Poaceae beside a whole static forest, where only some *Macaranga* and *Hagenia* progress. This peak of Poaceae pollen that would be assimilated to the Medium Poaceae Period (MPP) in Ngushu (Kabonyi and Roche, 2015) precedes the last expansion of *Sinarundinaria* a bit. This episode could be identified as the effect of the recognized dry climatic phase in Rwanda, 500 years AD (Roche, 1996; Roche and Ntaganda, 1999; Roche *et al.,* 2015).

8.5 DISCUSSION

The palynological study of the Cishaka sedimentary sequence should be placed in a larger regional context of the Upper Quaternary palaeoenvironmental evolution of the Lake Kivu borders. The specificity of the research is reflected in the fact that it provides additional information on other investigations made on other sites in the Mount Kahuzi area (Musisi I, Musisi II and Ngushu) around 2200 m a.s.l. One can trace the environmental dynamics during approx. 35,000 years, especially during the Last Glacial Maximum (LGM) and the Tardiglacial periods. The chronology of events has been established on the basis of ^{14}C dating.

Around 35,860 ± 1600 years BP, a disparate forest occupied the land that evolves toward a heterogeneous meso-ombrophile group which corresponds to moderately hot and humid climate, that is, the so-called "Kalambo interstadial" period (32,000–26,000 years BP, Clark and van Zinderen Bakker, 1964), subsequently followed by the beginning of the regressive phase of the "Mount Kenya Hypothermal" (25,000–15,000 years BP, Coetzee, 1967), including the Last Glacial Maximum (LGM).

During LGM, in the sector of the Kahuzi-Biéga, an exceptional expansion of herbaceous opened spaces is noticed. This reveals a development of natural high altitudes grasslands of dominant Poaceae, a phenomenon that is recognized when the dry and cold climatic conditions are installed. As it appears little likely that the eastern face of the dorsal of Congo could have sheltered some refuge forests during the LGM, these forests would have been located in small valleys sheltered in the same chain of the Mitumba as well as on the western face of the given hilly topography.

Later on, the environment becomes a forest again by the dominance of *Podocarpus*, associated with the Afro-subalpine elements, attests conditions remaining cold but becoming more humid, which justifies the expansion of the swampy flora. This event can be assimilated to the Tardiglacial.

In the forest resumption of the terminal Pleistocene, a new extension phase of natural grasslands occurs. Although, it is less important than the previous one, but it reveals a drought period very well. This phase can be considered as the contemporary of the Younger Dryas.

Concerning the Holocene period, one can find the same evolutionary process as for other sites within the Kahuzi area, e.g. Musisi I, Musisi II, and Ngushu (Moscol and Roche 1997; Runge, 2001a, Kabonyi 2007; Kabonyi and Roche, 2015).

During the Holocene Humid Optimum (HHO) occured in the Kivu between 10,000 years and 7000 years BP, the site did not achieve any environmental information because of growth of precipitation allowing a lacustrine evolution (Boutakoff, 1939; Runge, 2001a, 2001b). It is only from 7000 years BP, with the attenuation of the humid phase, extended to 5000 years BP that the swamp began to develop a new pollinic recording.

By contrast, on the eastern ridge in Rwanda, peaty sedimentation persisted, revealing the evolution of an Afromontane forest, from a heterogeneous status to a typical dense one that lasted from its maximum extension until ca. 7000–6000 years BP (Roche *et al.*, 2015).

The important growth of Poaceae in the pollen spectra, identified as the Older Poaceae Period (OPP, Runge, 2001a) and dated from 4000 years BP newly indicates an appreciable natural grasslands expansion in this altitude being accompanied by a "savanna" climatic development in the border of the Kivu. In the absence of an effective recognized human settlement, this progression of the opened savannas areas must, consequently, be considered as necessarily natural. The event would be due to a sensible weakening of the local hydrologic regime to put in parallel with the global phenomenon of aridity that hit the whole of Africa at this time.

Despite a forestry renewal observed since 3500 years BP, a climatic instability persists till ca. 2000 years BP. During that period, the increase of Bamboos areas occurs into the Afromontane forest. In the early first millennium of our era, after a return to a "normal" forest environment, a new peak of Poaceae was detected at the upper part of the diagram, just before a new progression of Bamboos. This episode, the so-called "Medium Poaceae Period" (MPP, Kabonyi and Roche, 2015), could be identified as the effect of the recognized dry climatic phase in Rwanda 500 years AD (Roche *et al.*, 2015).

In Cishaka, there is no posterior information related to this period because the sampling of the superior part of the sequence has been made aleatory by the fluidity of the organic matter in decomposition, full of water.

8.6 CONCLUSION

The Cishaka sequence, in addition to the confirmation of what is already known concerning the environmental evolution of the western ridge highlands, provides details about the last 20,000 years of the Upper Pleistocene period in the Kivu area. On the full sequence the following environmental phases were recognized:

- the Kalambo Interstadial (32,000–26,000 years BP) with the structuring of a heterogeneous Afromontane forest
- the Mount Kenya Hypothermal, including the LGM (25,000–15,000 years BP) which gave rise to an open landscape
- the Tardiglacial period with the forest renewal
- the dryness of latest Dryas and the development of open spaces
- the early Holocene lacustrine extension (10,000–7000 years BP) revealed by a thin sandy deposit and a pollen hiatus
- an Afromontane forest optimum around 7000–6000 years BP
- the extreme African aridity ca. 4000 years BP emphasized by the expansion of open grasslands
- a climatic instability around 3500–2000 years BP suggested by a progression of bamboos areas
- a forest renewal ca. 2000–1500 years BP similar to the present
- a short dryness ca. 500 years AD underlined by the MPP (Medium Poaceae Period) peak

The Upper Holocene events, among others the human impact on forest environment, were not recorded in the Cishaka core.

REFERENCES

Boutakoff, N., 1939, Géologie des territoires situés à l'Ouest et au Nord-Ouest du fossé tectonique du Kivu. *Mémoires de l'Institut Géologique de l'Université de Louvain*, **9**, pp. 1–207.

Clark, J.D. and van Zinderen Bakker, E.M., 1964, Prehistoric cultures and Pleistocene vegetation at the Kalambo Falls, Northern Rhodesia. *Nature*, **210**, pp. 971–975.

Coetzee, J.A., 1967, Pollen analytical studies in East and Southern Africa. *Palaeoecology of Africa*, **3**, pp. 1–146.

Combe, J., 1977, *Guide des principales essences de la forêt de montagne du Rwanda. MINAGRI* – Direction des Eaux et Forêts, Kibuye—Co-opération technique Suisse, Berne, p. 239.

Erdtman, G., 1960, The Acetolysis Method, a revised description. *Swank Botanist, Tidskrift*, **54**, pp. 561–564.

Fischer, E., 1996, *Die Vegetation des Parc National de Kahuzi-Biéga, Süd-Kivu, Zaïre*, (Stuttgart: Franz Steiner Verlag).

Habiyaremye, F.X., 1997, Etude Phytocoenologique de la dorsale orientale du Lac Kivu (Rwanda). *Annales Sc. Economiques*, Musée Royal de l'Afrique centrale, **24**, pp. 1–276.

Kabonyi Nzabandora, C., 2007, Etude palynologique de la séquence sédimentaire de Musisi–Karashoma II, Sud–Kivu (R.D. Congo) – Synthèse de l'évolution environnementale du Sud–Kivu au cours des deux derniers millénaires. *Geo-Eco-Trop*, **31**(1–2), pp. 147–170.

Kabonyi Nzabandoea, C., Salmon, M. and Roche, E., 2011, Le Parc National de Kahuzi-Biéga (R.D. Congo), patrimoine en péril?, Le secteur «Haute Altitude», situation et perspectives. *Geo-Eco-Trop*, **35**, pp. 1–8.

Kabonyi Nzabandora, C. and Roche, E., 2015, Six millénaires d'évolution environnementale sur la dorsale occidentale du Lac Kivu au Mont Kahuzi (R.D. Congo). Analyse palynologique de la séquence sédimentaire de Ngushu. *Geo-Eco-Trop*, **39**(1), pp. 1–26.

Lebrun, J., 1935, Les essences forestières du Congo Belge: Les essences forestières des régions montagneuses du Kivu, *I.N.E.A.C. Sér. Scientifique*, **1**, pp. 1–264.

Lebrun, J., 1936, *Répartition de la forêt équatoriale et des formations végétales limitrophes, Ministère des Colonies*, (Bruxelles: Publ. Direction générale Agriculture et Elevage).

Lebrun, J. and Gilbert, G., 1954, Une classification écologique des forêts du Congo, *I.N.E.A.C.. Série scientifique*, **63**, pp. 1–189.

Marchant, R. and Taylor, D., 1998, Dynamics of montane forest in central Africa during the late Holocene: a pollen-based record from western Uganda. *Holocene*, **8**(4), pp. 375–381.

Moscol-Olivera, M. and Roche, E., 1997, Analyse palynologique d'une séquence sédimentaire Holocène à Musisi-Karashoma (Kivu, R.D. Congo), Influences climatiques et anthropiques sur l'environnement. *Geo-Eco-Trop*, **1–4**, pp. 1–26.

Ntaganda, C., 1991, *Paléoenvironnements et paléoclimats du Quaternaire supérieur au Rwanda par l'analyse palynologique des dépôts superficiels*, Thèse de doctorat en Sc. Botaniques, Université de Liège.

Roche, E. 1996, L'influence anthropique sur l'environnement à l'Age du fer dans le Rwanda ancien. *Geo-Eco-Trop*, **20**(1–4), pp. 73–89.

Roche, E. and Ntaganda, C., 1999, Analyse palynologique de séquence sédimentaire Kiguhu II (Région des Birunga, Rwanda). *Geo-Eco-Trop*, **22**, pp. 71–82.

Roche, E., Kabonyi Nzabandora, C. and Ntaganda, C., 2015, Aperçu de la phytodynamique Holocène du milieu montagnard sur la chaîne volcanique des Virunga (Nord du Rwanda). *Geo-Eco-Trop*, **39**(1), pp. 27–54.

Runge, J., 1997, Alterstellung und paläoklimatische Interpretation von Decksedimenten, Steinlagen (stone-lines) und Verwitterungsbildungen in Ostzaire (Zentralafrika). *Geoökodynamik*, **18**, pp. 91–108.

Runge, J., 2001a, Landschaftsgenese und Paläoklima in Zentralafrika. *Relief, Boden, Paläoklima*, **17**, pp. 1–294. Gebr. Borntraeger. Berlin; Stuttgart.

Runge, J., 2001b, On the age of stone-lines and hillwash sediments in the Eastern Congo basin—palaeoenvironmental implications. *Palaeoecology of Africa*, **27**, pp. 19–36.

CHAPTER 9

The coastal Holocene sedimentary environments of Loango Bay and Pointe-Noire, Congo: Previous works, recent development and synthesis

Dieudonné Malounguila-Nganga
Département de Géologie, Université Marien Ngouabi, Brazzaville, Congo

Pierre Giresse
Centre de Formation et de Recherches sur les Environnements Méditerranéens, Université de Perpignan, France

Timothée Miyouna & Florent Boudzoumou
Département de Géologie, Université Marien Ngouabi, Brazzaville, Congo

ABSTRACT: This review focuses on Holocene swampy, estuarine and fluvial systems of the southern coast of Congo. These deposits can provide proxy data at various geographic and chronologic palaeoclimate scales. During the relative high stand of the MIS 3, about 45,000–35,000 years BP, an important thickness of peat deposited in mangrove swamp in the palaeo-valley of the Kouilou estuary. During MIS 2, the significant fall in the sea level led to the emersion of this site and to the important compaction of these peaty deposits. During MIS 1, the approach of the Holocene led new conditions favourable to the mangrove development. In the Bay of Pointe-Noire, the morpho-structural depression of the Upper Cretaceous between 9000 and 7000 years BP allowed the rapid deposition of more than 10 metres of not compacted peat. In Kouilou River palaeo-valley, it was only during the end of the Holocene that the bank deposition shows recurrent evidence of the mangrove proximity.

Holocene palaeoclimatic changes of central Africa are now well documented, but there is evidence that landscape evolution of the coastal plain has also been affected by approach and then arrival of the marine transgression and its influence on hydrological process. From 7000–6000 years BP, the construction of large beach barriers by the oceanic drift allowed the definition of long and narrow depressions where *Monopetalanthus* swampy forests developed, thus allowing accumulation of peat and organic muds. Water table oscillations favoured an active podzol process of the littoral sands and development of hydromorphic soils (gley). Indications of a marked renewal of erosion, probably of climatic origin, are present on the scale of the last centuries. Finally, a more recent generation of beach barriers came to bury these paralic deposits. Today, very dynamic oceanic erosion is exposing these Holocene deposits and providing new opportunities to obtain more and better palaeoclimate databases.

9.1 INTRODUCTION

In recent years, various lakes and swamps have been studied repeatedly, providing comprehensive reconstructions of tropical rainforest changes over the last millennia, throughout central Africa (Vincens *et al.*, 2000) and even beyond (Maley, 1997, 2012; Bostoen *et al.*, 2015). In southern Congo, there is no evidence of Quaternary marine deposits above the current oceanic zero. The Holocene sea reached the current level approximately 6000 years ago and has never overtaken it since (Delibrias *et al.*, 1973; Giresse *et al.*, 1984).

During the last few decades, various Holocene swampy soils and deposits of the Congolese coast were studied either in an outcropping position, such as in Loango and Pointe-Indienne, or buried under recent sands of the offshore bar, as in the Kouilou and Songololo estuaries (Delibrias *et al.*, 1973; Giresse and Kouyoumontzakis, 1974; Giresse and Moguedet, 1982; Schwartz, 1985; Dechamps *et al.*, 1988; Elenga *et al.*, 1992; Elenga *et al.*, 1996). Presently, under the action of oceanic erosion, numerous new outcrops have appeared especially in Loango Bay. Such occurrences provide an approach for a new and strongly widened Holocene history of the continental environments close to the shoreline.

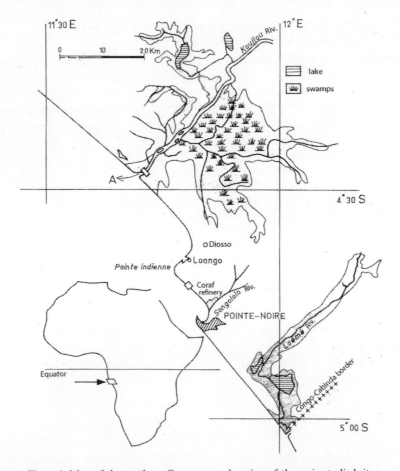

Figure 1. Map of the southern Congo coast, location of the main studied sites.

Various older studies related to the Quaternary marine and/or palustrine buried deposits were published in a fragmentary way and in small-circulation magazines. Our intent is to collect these data to make a critical analysis and to propose some new synthetic consideration. Several new dates and new observations are also included within the framework of this development. Going from north to south, the most significant sites shown in Figure 1 are Kouilou Estuary, Diosso Circus, Loango Bay, Coral Refinery, Songololo River, and Pointe-Noire Bay. Lake Kitina, at some 40 km from the coast, is out of the map.

9.2 STUDY SITE

The equatorial West African continental margin comprises depot centres located off-shore major river mouths (e.g. Ogooué, Congo, Cunene, etc.). Accumulation rate in these deep-sea-fans steadily increases from Early Oligocene, and reaches a maximum in the Plio-Quaternary time when Plio-Pleistocene of the "Série des Cirques" (amphi-theatre-like erosional forms) accumulated. Such increase in terrigenous sedimentation indicates an increase of continental erosion in the corresponding watersheds throughout Cenozoic (Séranne *et al.*, 2011). A 1000-km-long N-S trending range, the Mayombe mountain range, separates the Congo Basin from the western flank of the basin. Seaward of these mountains, Cenozoic non-marine sedimentary rocks occur on the lower slope of the African Surface Basin. The 3 to 6 km wide coastal plain of Congo presents a superficial ochre sands cover that is the result of recurrent reworking of sands of the piedmont of the "Série des Cirques". This cover can locally be colluvial or alluvial, but recent works on the nearby area of Gabon (Thièblemont, 2012) also show the wide importance of the aeolian processes over the three last millennia.

This coast is affected by a marked dry season of 4 to 5 months' duration because it is submitted to the linked influence of the trade winds blowing from the southeast and of the cold Benguela current. The annual rainfall is presently lower than 1300 mm and the average temperatures range from 22° to 25°C. The vegetal cover is generally constituted by forest-savanna mosaic, with low savanna with *Ludetia arundinacea* and with swampy forests with *Symphonia globulifera* along rivers and in freshwater ponds between the beach barriers, where herbaceous meadows of *Cyperus papyrus* are also widespread. The mangrove swamp with *Rhizophora* is quite rare and restricted to tidal zones of the Kouilou estuary where it can penetrate several kilometres upstream. Soils are commonly ferrallitic and psammitic with up to 95% sand content; while hydromor-phic soils and podzols are restricted to the wet areas. The podzolic upper horizon is strongly leached and its white colour characterizes the wide surface of a littoral band (Jamet and Rieffel, 1975). The real age of the beginning of the podzolization process could be of the order of 3000–3500 years BP (Schwartz, 1985; Elenga *et al.*, 1992).

9.3 THE ESTUARINE VALLEY OF THE KOUILOU RIVER

The estuarine site of Kouilou River is the best documented on the shoreline of Congo regarding palaeoenvironmental reconstruction at the scale of the last 40,000 to 50,000 years. Two successive projects of building or renovation of a road bridge required the drillings of deep geotechnical soundings. In 1977, in a first step of sur-vey of the rock basement, the drillings did not generally exceed thirty metres deep and have only rarely reach the top of the Upper Cretaceous siltstones, that are probably Senonian in age (Giresse and Moguedet, 1982; Moguedet *et al.*, 1986). In

1984, the second drilling phase allowed recording of a 50 m deep set of cores, which penetrate in the top of the Upper Cretaceous (Malounguila-Nganga *et al.*, 1990). These sections supplied by thirteen soundings, supported by fourteen radiocarbon datings, allowed the Late Quaternary reconstruction of a wide part of the Kouilou palaeo-valley (Figures 2a and 3).

The surface of the Upper Cretaceous appears very uneven because nearly 20 m difference in level were observed on less 100 m of distance. A rather narrow channel, more than 40 m deep, is more or less under the current channel; while another depression also deeper than 40 m lies under the current right bank of the estuary, which suggest a more extended northern main valley that could correspond to an older mouth of the river (Figure 2a).

9.3.1 Deposits of mangrove swamps older than 35,000 years BP

The first sedimentary unit (Unit 1) is constituted by very compacted black organic clays (average 5% organic Carbon), which were completely penetrated in spite of a >15 m thickness. Millimetre-thick or centimetre-thick quartz layers are interstratified in the lower part of this jarosite-rich organic deposit; in the upper part, the jarosite decreases, whereas the clay content increases. Unit 1 is recognized only below the present right bank where it extends over ca. 300 m width. A radiocarbon date gave an age beyond 35,000 years BP, and two others carbon dates provided of 46,517 years BP and 36,124 years BP, which are stratigraphically coherent. These imply, after compaction, a sediment accumulation rate

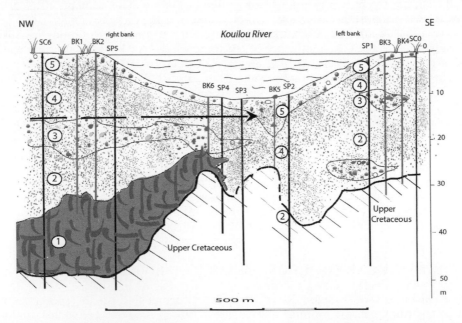

Figure 2a. Stratigraphy of the Late Quaternary riverine-marine filling of the Kouilou estuary above the top of the Upper Cretaceous. According to the location of the soundings, correlation of the sedimentary units 1–5. Unit 1: Deposits of mangrove swamps previous to 35,000 years BP, Unit 2: Deposits of estuarine banks between 9000 and 5000 years BP, Unit 3: Deposits of alluvial channels around 5000 years BP, Unit 4: High stand estuarine deposits, and Unit 5: Recent alluvial deposits. The arrow indicates of the main channel during the filling.

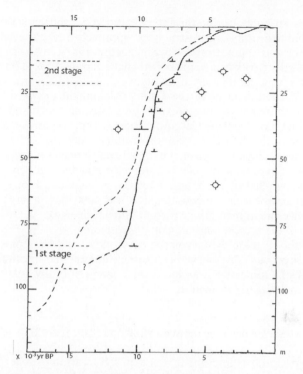

Figure 2b. Relative sea level curves of South Gabon and Congo area (solid curve) and of Ivory Coast area (dashed line). The vertical bar indicates that its top corresponds to the mean sea level, the horizontal bars represent the range in radiocarbon dating. Circle points are excluded from the curves (after Giresse *et al.*, 1984).

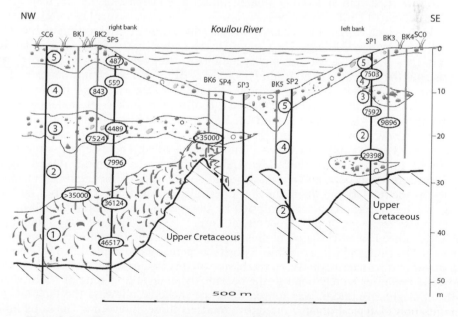

Figure 3. Distribution of radiocarbon ages compared to the successive sedimentary units.

of 1.21 m/10³ years. Probably, during this interval, the MIS 3 shoreline was slightly close, its height according to isotopic reconstructions approached 50 m, even 30 m (Imbrie *et al.*, 1984; Chappell and Shackleton, 1986; Shackleton, 1987). In any case, the deepness of Unit 1 rules out MIS 5 evidence. Some pollen analyses were realized on the SP5 sounding and indicate a strong dominance of mangrove pollen and a warm and wet climate particularly in the deepest deposits of this Unit 1 (Malounguila *et al.*, 1990). According to the retreat of the shoreline, the mangrove swamps became scarce and made way for a more continental and more confined swampy environment. In parallel, a growing aridification process is indicated by the increase of savanna pollen (up to 20%). During MIS 3, the slightly organic sediments deposited in an alluvial depression that is different from the current channel. No organic deposit is observed below the right riverbank, possibly because of the uneven relief of the Upper Cretaceous basement. However, other channels functional further north can be considered. It was most likely a large estuarine area, which was more widespread than it is at present and which evokes the marked extension of swampy waters of the present coastal plain of Kouilou.

This site of the Kouilou estuary is the only one of the Congolese coast, where MIS 3 deposits are proven. They have been sometimes implied in other points of the coast of the Gulf of Guinea such as in Sierra Leone (Anthony, 1983), Ivory Coast (Fredoux, 1980), and Nigeria (Sowunmi, 1981).

9.3.2 Deposits of estuarine margins between 9000 and 5000 years BP

The top surface of Unit 1 is uneven. It was exposed to a long subaerial exposure, in particular, during the prolonged low stand of the MIS 2 inducing the sediment compaction. It is likely that the depression was flooded by the active flow of Kouilou waters, which levelled the top of the formation. Evidences of this alluvial period are scarce, if it is not some 2 to 4 m-thick sandy beds observed in SP5 and SP2, where an organic layer probably reworked from underlying organic muds was dated in 29,398 years BP.

The deposit of Unit 1, which peaks towards 20–25 m, was covered by the Holocene sea level rise which happened around 9000 years BP leading to the setting of the second sedimentary unit (Unit 2). The Congo Holocene sea level curve is slightly above the global sea level curve without any specific (regional) characteristic. The oceanic level reaches the current zero at around 5500 years BP, a level which it never exceeded afterwards, despite some slight oscillations (Figure 2b) (Delibrias *et al,* 1973; Giresse *et al.*, 1984). At this time, several clasts of the roof were reworked and redeposited in the first accumulation as observed in BK1, BK2, SC6, and SP1. On the higher points where the organic muds were absent, the Holocene deposits were directly transgressive on the Upper Cretaceous whose top was eroded and levelled. These deposits are rather well-sorted fine sands. Under the left bank, dark pelitic lenses are frequently interbedded. The more sandy deposits suggest deposition in the axis of the depression widely invaded by the tidal stream where the flow controlled the upstream transfer of marine particles. The marine character is attested by the presence of abundant small faecal pellets in the first steps of the glauconitization process, also evidenced by sometimes strongly oxidized small clasts of mollusc shells and benthic foraminifera (*Ammonia beccarii*). The finer and more organic deposits, even mangrove muds dated 9896 years BP correspond to the proximity of the banks of this estuarine valley. However, this landscape remained largely open to the oceanic dynamic and in spite of the slowing of the transgressive rate, it was not confined or sheltered allowing the long-lasting accumulation of typical organic mangrove muds as those observed at the same time, for example, in the subsoil of Pointe-Noire (Giresse and Kouyoumontzakis, 1971).

9.3.3 Stream channel deposits towards 5000 years BP

The following interval constituting sedimentary Unit 3 corresponds to the peak of the Holocene transgression. It was dated only once and its chronology is thus considered according to the other underlying or overlying dated deposits. It is a layered accumulation of coarse riverine gravels, even quartz and sandstone pebbles and lateritic scoria. It can reach until 6 m of thickness in BK1 and 8 m in BK2 and corresponds to the alluvial lens of old main channel of more than 400 m wide; but the current soundings informed us only about its southern part which bevels southward (Figure 2). This transport episode characterizes a channel location very different from that of today. It also corresponds to the last time of wet period from 6000 to 3000 years BP as shown by the study of the nearby Lake Kitina (Bertaux *et al.*, 2000; Vincens *et al.*, 2000) when the flow of the river was still probably at its higher energy. Above, the marine indicators (green pellets, bioclasts) become scarce. Strongly oxidized and decayed quartz indicate the reworking of the Pleistocene piedmont of the "Série des Cirques" (Giresse and Le Ribault, 1980). After the deposition of Unit 3, the filling of the valley was largely carried out (about two-thirds), from the Mid-Holocene, the accommodation space was restricted and the estuarine landscape was becoming more stable.

9.3.4 Estuarine high stand deposits

The high sea level being reached, the facies transition from low energy marine swamps in the north to lenses of sandy alluviums in the south is recognizable. The main channel tends to migrate to the south of the high Cretaceous basements. From this point, the sedimentary unit (Unit 4) shows a rather clear contrast between:

a. the northern area of the palaeo-valley where the end of the filling is realized with fine or sometimes clayey and organic sediments, which corresponds to the last millennium (843, 559 and 487 years BP), while other deposits attest the closeness of the mangrove vegetation on the banks (oyster clasts) and a marked oceanic influence (thin *Mactra* shells); and

b. the southern part, where deposits are coarser and irregularly sorted including a hundred metres-wide lenses of sands and gravels. Several deposits between 15 and 5 m deep are dated between 9000 and 7000 years BP and Unit 4 is considerably reduced, even absent.

In the course of its sedimentary history, this alluvial valley preserved nearly no alluvial or estuarine sedimentary evidences for the interval between 4500 and 1000 years BP. Successive erosion of these exposures can be related to avulsions of the main and secondary channels, but also to the general downstream progradation of the alluvial deposits in the period of oceanic level stability. A climatic interval with globally more contrasted and abrupt floods, and thus more erosive, can be envisaged.

9.3.5 Recent alluvial deposits

Through all the width of the estuarine valley, the cover of the recent alluvial deposits of the Kouilou River can be followed (Unit 5). The thickness of this sedimentary unit ranges from 2 to 5 m except below the main channel where it reaches about 10 m (Figure 3). The medium sands are slightly sorted, finer near banks, and similar to those of the current alluviums of the river. Some glauconitized faecal pellets still testify in the favour of estuarine flow. The overflowing of the floods ends in the emersion of a part of the banks where hydromorphic soils (gley) develop as well as in the growing narrowness of intertidal surfaces where the mangrove often almost disappears.

9.4 PALAEOCLIMATIC RECORDS OF LAKE KITINA

The sedimentary archives of the lakes of Cameroon, Gabon and Congo supply the best Holocene biostratigraphic and palaeoclimatic information. Within the framework of the evolution study of the coastal region of Loango, studies of Lake Kitina, located on the western border of the Massif of Mayombe at approximately 40 km from the shoreline, constitute the most appropriate references (Elenga *et al.*, 1996; Bertaux *et al.*, 2000).

It is a lake or a deep swamp, 4–5 m deep, 5 km long and 1–2 km wide, which partly occupies the palaeo-valley axis which incise the Precambrian metamorphic basement of the Mayombe Massif (Figure 4). A 6.25 m-long core provided the palaeoclimatic evolution for approximately 5400 years long that is 6250 cal years BP (Figure 5a, b). The sediment core shows between 6.25 and 2.10 m deep rather homogeneous grey

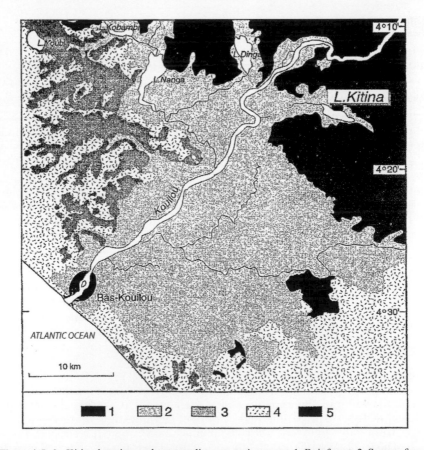

Figure 4. Lake Kitina location and surrounding vegetation cover. 1. Rainforest, 2. Swamp forest, 3. Coastal mesophyll forest, 4. Savanna, 5. Mangrove (after Elenga *et al.*, 1996).

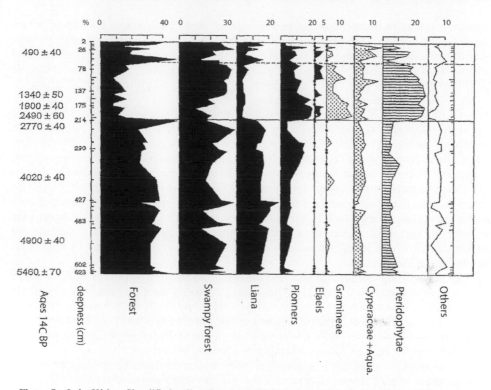

Figure 5a. Lake Kitina. Simplified pollen diagram KT3 from Lake Kitina (south-western Mayombe) (after Elenga *et al.*, 1996).

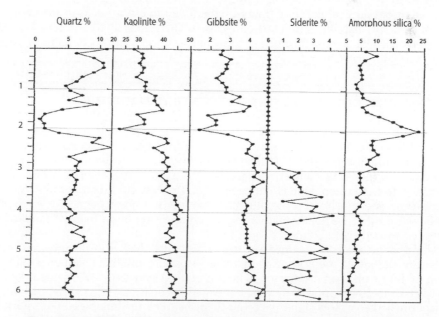

Figure 5b. Lake Kitina. KT3 main minerals observed by the FTIR spectroscopy method (after Elenga *et al.*, 1996).

organic muds with frequent vivianite and siderite nodules; between 2.10 and 0.60 m, the deposit is more organic and the vivianite and the siderite tend to disappear. Finally, near the top section, the deposit is a dark grey sapropelic mud. Main detrital minerals are quartz, kaolinite and, additionally, gibbsite, but between 2.10 and 1.60 m (between 2490 and 1340 years BP), their accumulation was strongly diluted by the high abundance of the amorphous silica of diatoms.

From the pollen analyses, between 5460 and 2490 years BP, the forest elements represent 60 to 80% of the pollen sum; the other taxa correspond to Cyperaceae and to various lianas associated to Pteridophyte spores. Between 2490 and 490 years BP, where the diatoms episode is located, we identify a major change of the surrounding vegetation, wherein non-swampy forest taxa decreases clearly for the benefit of heliophilous pioneers, Pteridophyte and especially Graminae which peak at 50%. Precipitation and detrital flows decreases are likely; the rain-wash efficiency always in forest dominant setting, being slightly reduced. The planktonic diatom presence shows that the lake did not dry out even if its level probably fell. The last 500 years indicate the progressive return of forest, lianas, and Cyperacea taxa to the detriment of heliophilous and herbaceous taxa.

In general, the immediate nearness of the great Mayombe rainforest confers on this site a special connotation in that the trees cover was almost permanent on the scale of this record. Even during the driest periods, the conditions remained favourable to the forest cover. However, in spite of this general character, it was possible to observe environmental changes even if they were less intense and took place later than in the less stable nearby regions inhabited today by forested savanna.

In summary, we identify a phase of very late recolonization associated with a clear increase of the precipitation after 500 years BP, where included savannas ("savane incluse") disappear. A similar episode was also recorded in Niari plateau at about 650 years BP (Vincens *et al.*, 1994) and towards 600–550 BP on the nearby Atlantic coast (Schwartz *et al.*, 1990; Elenga *et al.*, 1992). These dates can be moved closer to the rather recent ages of the final filling of the Kouilou estuary and to those of the alluvium or colluvium deposition at Loango Bay. There is likely to be a link between these trends without, for the time being, being able to specify their geographical extension or their edaphic meaning.

9.5 SUB-SURFACE DEPOSITS OF POINTE-NOIRE HARBOUR

The Pointe-Noire Bay does not present real river mouth, only the depression of Tchikoba lagoons could show remains of a former drainage axis diverted from Loeme River during the last low stand. Unlike sub-surface of the Kouilou mouth where a river incision in the Cretaceous extends to 50 m deep, the depressions of Pointe-Noire sub-surface do not exceed 30 m (Figures 6a, 6b). The NW-SE orientation of the current coastline is controlled by the morphostructure of Cenomanian, Turonian and Senonian strata that are lightly undulating (dips are always lower than 10°) and often interrupted by orthogonal faults.

The seismic records obtained on the continental shelf also demonstrate light undulations that affect the Eocene and Cretaceous strata and that favoured a selective erosion of marls and some sandstone (Jansen *et al.*, 1984). NE-SW oriented depressions with 10 to 20 m topography were excavated in the top of the Cretaceous marls. They channelled some low stand river and then sheltered the first lagoon or intertidal depositions of the Holocene transgression. It was precisely off Pointe-Noire that 10 m isopach of loose sediments was registered (Jansen *et al.*, 1984).

The enlargement of the port gave opportunity for more than 200 soundings, which supplied an exceptional collection of documents about the sub-surface deposits (Giresse and Kouyoumontzakis, 1971). In most of these sub-surface sections, we can distinguish three main sedimentary units (Figures 6a, 6b).

a. the first unit (Unit 1) is composed of 2–5 m-thick azoic lower sands distributed along two channels. However, this unit is locally absent at the approach of the basement highs. There are white podzolised sands drained from the littoral plain during the last low stand. On the slopes, some coarse oxidized remains of the "Série des Cirques" are still preserved.

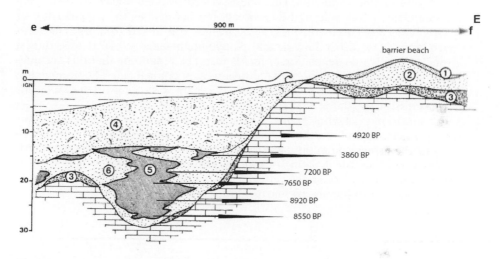

Figure 6a. Section of the Pointe-Noire subsoil perpendicular to the shore line (SW-NE). 1. Leached white sands deposited in channels, 2. Peats of mangrove swamps (7000–8000 years BP), 3. Holocene shelly sands, 4. Coarse pediments of the "Série des Cirques", 5. Ochreous sands on beach barriers, and 6. White leached sands (podzol).

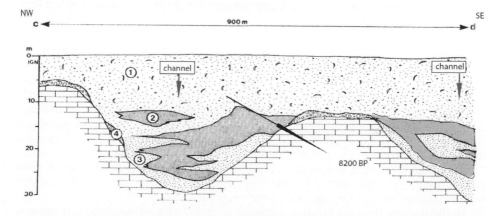

Figure 6b. Section of the Pointe-Noire subsoil parallel to the shoreline (NW-SE). 1. Leached white sands deposited in channels, 2. Peats of mangrove swamps (7000–8000 years BP), 3. Holocene shelly sands, and 4. Coarse pediments of the "Série des Cirques".

b. Unit 2 is constituted by more or less silty or clayey peaty accumulations developed in the depressions where they extend beyond lower sands. This unit can involve a single 2–5 m-thick accumulation or recurrent beds with fine sand interbeds. The higher thickness of this unit can reach about 12 m. The remains of *Rhizophora* (trunks, roots, and seeds) and the oyster shells debris indicate an important mangrove swamp accumulation growing in intertidal areas sheltered from the main dynamic of estuary waters, which is a markedly different environment from those of Kouilou. Seven carbon datings were obtained leading to ages between 8920 and 7200 years BP, with an only exception of 3860 years BP probably connected to a contamination from overlying marine sediments. The oldest ages correspond logically to the deepest deposits between 25 and 29 m deep, but because of the irregularities of the Cretaceous basement, these ages are not still linked to altitudinal steps (Delibrias *et al.*, 1973; Giresse and Kouyoumontzakis, 1974). It was, thus, a very active organic deposition (7 m/10^3 years) by reminding that this accumulation knew no emersion and is always weakly compacted. The average age of these accumulations coincides with the marked slowing of the Holocene transgression estimated near 8000 years BP (Giresse, 1981).

c. The 4 to 10 m-thick Unit 3 consists of shelly sands rich in *Anadara senilis, Ostrea denticulate,* and in other heterodont bivalves. It expresses the final flooding of the depression upon the arrival of the Holocene transgression towards 7000–6000 years BP. A dating of mollusc shells supplied a 4920 years BP. The increasing muddy accumulation achieves the filling of the depression before the overflow into the nearby marine bottoms.

9.6 LOANGO BAY

Loango Bay shows a low coast lined with littoral barriers except near Pointe Indienne, where the "Série des Cirques" cliff approach the coastline. These barriers constitute two, even three alignments. There are wind accumulations stemming from the deflation of the beach, which were fed on a large scale by a powerful littoral drift. These barriers, which can achieve until 10–15 m height, follow one another on the coast and express a seaward high stand progradation (Giresse and Kouyoumontzakis, 1974). The barrier deposits as those of the coastal plain are generally constituted by ochre sands, sometimes overlying directly on coarse and oxidized colluvium from the piedmont of the Series des Cirques. Ochre sands, generally well-sorted and rather rich in dull quartz grain, resulted from various re-workings of the piedmont after several aeolian, colluvial, or alluvial episodes. The upper deposits were the result of an active podzolization process ending in the definition of a superficial horizon A of leached white sand of some 0.5–1 m thickness (Schwartz *et al.*, 1992) and of a brown accumulation horizon B rich in organic matter and iron oxides with discrete alios concretions. This brown, sometimes dark brown horizon when it is hardened, may contain remains of fire places (charcoals), potteries, or even Neolithic industries (Tshitolian). From about 1970, this horizon was largely exposed to the active oceanic erosion, in particular near the Catholic Mission where it was once confused with a peat deposit. Charcoals sampled in the podzolic brown horizon close to the CORAF refinery installations supplied an age of 1890 years BP.

In the last few decades, the deterioration of erosion has allowed the exposure of further Holocene paralic deposits, which have been subject to new studies.

Organic matters of both unconsolidated and hardened brown horizons located near Djéno Rocher were ^{14}C-dated up to 1330 and 1880 years BP, respectively, that apparently corroborated the age previously obtained in Pointe Indienne (Schwartz, 1985). However, the renewal of the organic matter during the accumulation process lead to the integration of probably successive organic accumulations that tends to modify the measured ages. The real age of the beginning of the podzolization process could be of the order of the double, i.e. 3000–3500 years BP, going back at the beginning of the climatic drier episode in Central Africa (Elenga *et al.*, 1992). This episode was one of the first that caused wind-borne accumulation of ochre sands (Thièblemont, 2012). Little before 1980, a geotechnical study was realized on the future place construction of a new oil refinery, i.e. 10 km North of Pointe-Noire. This site, located a little behind the coastal barriers, was the object of some 40 preliminary soundings, from which it was possible to examine nearly half of the soundings and to analyse in detail three of them (Giresse and Moguedet, 1980). The Cretaceous was reached every time allowing recognizing an only 10 m-thick Quaternary cover contrasting with the 30 m cover recorded in the Kouilou palaeo-valley, the greatest thickness corresponding to the littoral barriers. This cover consists of coarse and oxidized sands with some gravel beds in the lower part, near the top of the "Série des Cirques". Sands cleared by podzolization are exposed near the summit. In these deposits, there is no occurrence of organic rich soils or organic accumulations, which is sometimes scoured by the swell in the present upper beach.

To be complete, two radiocarbon dates were obtained from other samples (Giresse and Moguedet, 1982). At Loango, solid black clay was discovered at one metre below the present sea level contained an abundant fauna of small young shells of *Anadara senilis*. These shells were dated at 520 years BP providing the only past marine or brackish waters evidence of this coastline and showing the recent progradation. This was again followed by erosion of the Bay. On the banks of the Songololo River mouth, a small river coming out in the north of the Pointe-Noire Bay, some slabs of blue-grey or blackish claycy deposits appear at 0.5 m above the present high water. Their organic matter was dated at 2920 years BP. More or less organic hydromorphic soils (gleys) are found here, the age of which testifies of the last time of a swampy landscape through which we can follow the history in the nearby sites. These outcrops of restricted extension indicate, during the Mid-Holocene, the likely extent of swamps or ponds in the Pointe—Noire Bay possibly similar to those of the Loango Bay; but in this partly sheltered littoral, evidences are probably deeply buried under the beach sand. Until now, the most documented study of Holocene deposits in the northern area of Pointe-Noire is the one of Schwartz *et al.* (1990).

9.6.1 Site of Loango

It was the most continuous vast outcrop of the bay where on a nearly 200 m length, the incision of the swell cuts the 2–3 m-thick aliotic horizon B (Dechamps *et al.*, 1988). The remains are well-preserved roots, especially taproots are implanted in the alios which they cross vertically indicating an implementation previous to the compaction or the cementation of the horizon. A more irregular crossing could indicate the difficulties of implanting in a compacted soil (Figure 7). Today, these red to brown stubs of shrubs are spectacularly washed by successive uprushes. In spite of nearness to the ocean, the botanical analysis of 18 flora species show characteristics of trees of the rainforest, even of the swampy forest but without influence of brackish waters. Big trees with various species of *Monopetalanthus* and *Saccoglottis*, medium trees and shrubs were found at the same place. It was a dense hydromorphic forest affected successively by periods of dehydration and flooding and involving the oscillation of water

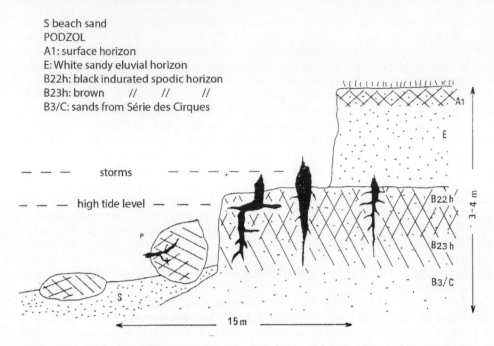

Figure 7. Eroded Loango upper beach. Schematic succession of the podzol horizons showing implementation of swampy forest. Roots and taproots of the great *Monopetalanthus* cross over all the podzol accumulation horizons (after Schwartz *et al.*, 1990).

table. This last one is especially favourable to the podzolization, while areas of stable water table can correspond to local accumulations of peat bog type that are observed here a little downward (Schwartz, 1985). In a general way, humus of the litters of the forest supply organic acid matters that are very favourable to the progress of the podzol. Five pieces of woods of *Monopetalanthus* were dated between 5800 and 3100 years BP, that is an interval corresponding to a still wet Holocene and which precedes the trend towards slightly drier conditions. Three other carbon datings of organic matters of the accumulation horizons of the podzol indicate ages between 6540 and 3700 years BP, confirming the contemporaneous of the whole vegetation cover.

9.6.2 CORAF Refinery

Two nearby sites of the CORAF refinery are located some 10 km south of the Loango site. These sites correspond to a shoreline sector sheltered from the direct attack of the swell and where outcrop occurrences are induced by height seasonal storms. There are still *in situ* roots and taproots included in podzol brown horizons or locally in peat accumulation (Figure 8). The florae of the rainforest and the swampy forest appear very similar to that of the Loango site. A wood and a peat dated respectively at 3740 years BP and at 4140 years BP indicate a synchronism with the Loango site. The base of a 60 cm-deep sediment core supplied an age at 3060 years BP, while the middle part and the top layer of the same section are respectively dated at 1590 years BP and contemporaneous. The pollen study permits a palaeoenvironmental outline, the conclusions of which join the main lines of the studies related to the nearby sectors (Elenga *et al.*,

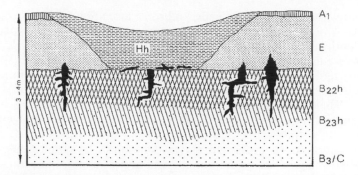

Figure 8. Section of CORAF site. Successive horizons sub-outcropping at the top of the beach (A1, E, B22h, B23h) above the ochreous sands of the coastal plain (B3/C). Here, the soil ends with peaty accumulation (Hh) (after Elenga *et al.*, 1992). Like in Loango, various roots are implanted in the brown soil horizon.

1992). Towards 3100–3000 years BP, the landscape remained widely forested. The littoral plain or at least a part of this one was occupied by a dense and periodically flooded rainforest with *Syzygium* that was long lasting, probably from the beginning of the Holocene. Such vegetal cover implied a littoral rainfall slightly higher than nowadays. Between 3100 and 1600 years BP, herbaceous taxa roughly felled from 60 to 10% and the opening of the cover increased, thus suggesting a forest-savanna mosaic. None of these organic deposits had been crossed by the numerous soundings preliminary to the refinery building, which indicates a rather narrow swampy depression setting under the present beach barrier and possibly more seaward. Around 1600 years BP, a brief forest recurrence is characterized by the presence of a *Syzygium* secondary forest, probably linked to the water table level of the swampy forest connected with that of the ocean. Thereafter, decrease and disappearance of the forest environment is observed, which were accompanied by the development of heliophilous taxa such as *Alchornea*. The pollen assemblage contains more than 80% of Gramineae associated with swampy taxa.

It appears that these forest and swamps occurrences during the end of the Holocene were not punctual features. Their lateral extent are evidenced on the coastline of more than 10 km long and it is likely that new discoveries will be effective in line with the present and future coastal erosion of the Congolese shore. At about 6000 years BP, the settling of seasonal or long-lasting swamps was certainly enhanced by the growing rainfall, but these precipitations were previously active since several millenniums already. It was at the same time the approach and the arrival of the shoreline to its present level were determined by decreasing or stopping the downward drainage of continental waters. The rapid accumulation of beach barriers linked to the powerful oceanic drift ensured the isolation and the settlement of the swampy ponds.

9.6.3 Diosso Circus

This site corresponds to the upper beach below the erosion amphitheatre of Diosso, where a 500 m-long reddish clay accumulation supplied from the amphitheatre strikes off. The deposit is partly affected by water-influenced soil types (gley) induced in a previous swamp environment. Some tree roots or even bases of trunks are still in life position. Locally, transported branch fragments and peaty accumulations are observed. Above this hydromorphic soil, 1 to 3 m-thick alluvial or colluvial deposits with recurrent sandy-clay or clayey-sand beds, and some more organic layers are interbedded (Figure 9). Woods pieces are generally of grey-white colour (white wood). Twelve taxa were identified, among which *Annona glabra*, *Alstonia congensis* and *Agelaea* are all essences either strictly or preferentially of wet environment. In spite of the immediate nearness of the shoreline, there is no indication of brackish influence. For the greater part, the same species are still present in the current outlet of the Diosso amphitheatre. Three ages obtained by ^{14}C range between 480 and 600 years BP are much younger than those of Loango. The most significant palaeoclimatic information stemming from this study concerned the transition from a fine grain deposition in swampy environment to a coarse grain alluvial deposition. This transition marked the trend of strengthening of the erosion of the amphitheatre and, thus, the increase in precipitation at least at the scale of this coastal region. The meaning of this signal will be considered in the more widened scale of the succession of the sediment accumulations at the west of the Mayombe Massif.

9.6.4 Songololo River

A 5.4 m-deep vibro-core drilling was realized on the swampy borders of this river in approximately 3.5 km of the shoreline and near Pointe-Noire (Elenga *et al.*, 2001). The studies of pollen, the mineralogical composition and the carbon isotopic ratio of the organic matter indicate that this sector of small valley was occupied from approximately 7000 to 4000 years BP by the swampy rainforest characterized here by *Hallea* and *Uapaca* and by a nearby mangrove swamp. It implies that, at some point in the past, the estuarine mouth had to communicate with the swampy hinterland allowing tidal salt-water penetration. The sandy-clayey deposits indicate a filling in slightly quiet environment, which started at the approach of the Holocene sea level. At about

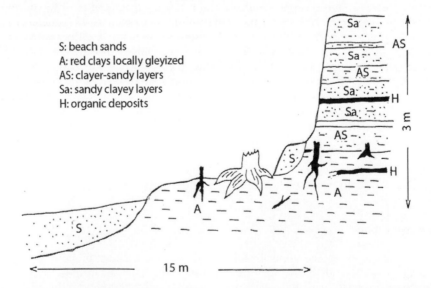

Figure 9. Diosso upper beach, outcrop of red clay deposited in freshwater swamp (after Schwartz *et al.*, 1990). The vegetal remains (roots, base of trunks) are implanted in the clay deposit, fragments of branch were transported over short distances. This clay deposits and its rests were buried under several sequences of riverine sands.

3000 years BP, the mangrove swamp disappeared and the environment evolved toward a swampy Cyperaceae and fern meadow out of reach of the ocean. This evolution was controlled more by the filling progress and by the accumulation of the beach barriers than by a hypothetical negative movement of the oceanic level. The increase in fern here indicates a transition towards a forest retreat. From 4500 years BP onwards, we record a decrease in the sediment accumulation rate in connection with a rainfall decrease. This trend was also recognized in other various sites of central Africa, i.e. forest extent between 7000 and 3000 years BP, subsequently it retreats after 3000 years BP. Here the swampy forest fragmentation process enhanced the spreading of the hygrophilous herbaceous plants dominated by Cyperaceae and *Raphia*.

This site of Songololo is original because it associates in close places two consequences of the arrival of the Holocene marine transgression: the flood of the depressions of the coast because of the rise of water tables and, doubtlessly, more locally, the establishment of a small estuary with its salt tide. The progress of the filling will make these landscapes disappear. After 3000 years BP, these processes were accompanied and enhanced by the climatic evolution.

9.7 MAIN RESULTS AND CONCLUSION

The Late Quaternary organic accumulation of the southern Congo coast was controlled first of all by the closeness of the oceanic water masses. It was logical that mangrove peats settled only near the shoreline when a structural morphology offered a shelter, even a favourable enclosed area.

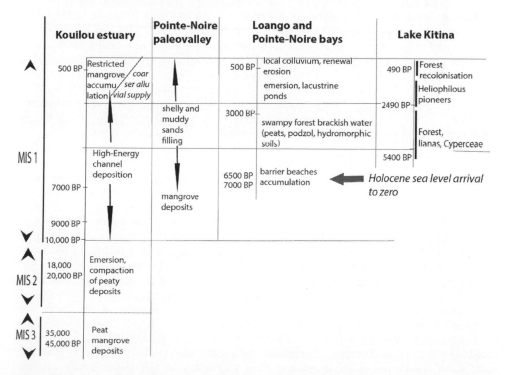

Figure 10. Chronological diagram of coastal evolution of main sites and Lake Kitina vegetal cover episodes.

The oldest Pleistocene deposits were discovered in the palaeo-valley of the Kouilou estuary, where at about 45,000–35,000 years BP, i.e. during the relative high stand of the MIS 3, an important thickness of peat deposited in mangrove swamp accumulated (Figure 10). This accumulation developed under cover or away from the destructive shifting of the main channel in a vast swampy area retained by the transgressive approach of oceanic waters. Then, the significant fall in the sea level (MIS 2) led to the emersion of the site and to the important compaction of these peaty deposits. Later during MIS 1, with the same effects as those of the previous high stand, the approach of the Holocene led new conditions favourable to the mangrove development. In the Pointe-Noire Bay, the morphostructural depression of the Upper Cretaceous allowed the rapid deposition of more than 10 m of non-compacted peat between 9000 and 7000 years BP. The cessation of this process was connected to the end of the marine transgression, which came to submerge the site and to bury it under several metres of shelly and muddy sands (Figure 10).

However, the same causes did not produce the same effect. In the valley of the Kouilou estuary, according to the avulsion of the main channel, the following filling of the valley was made under high energy hydro-sedimentary conditions different to those prevailing in a mangrove swamp. It was only towards the end of the Holocene filling that the marginal deposits above the current banks show recurrent evidence of the mangrove proximity.

A marked originality of this coastal sector of Congo lies in the wide extension of peaty swamps contemporary of the Holocene marine high stand, which are attacked and exposed by marine erosion today. Consequently, the most eroded coastal sectors allow the best observation of these Holocene deposits, but we can presume that similar deposits exist under the recent beach or beach barriers of preserved shoreline as that of the Pointe-Noire Bay. These deposits, without any salty or even brackish water influences, were dated between 6500 and 3000 years BP (Figure 10). They accumulated on several hundred metres-long depressions where long-lasting or seasonal swamps sheltered the growing of big trees of the swampy forest or dry land forest. These swamps, shielded by barrier beaches that have disappeared today, spread in connection with the rise of oceanic water level that blocked the downstream flow of freshwater table. The emersion intervened corresponding with the completion of the filling. The swell erosion has probably, partly cleaned the roof of these accumulations. Then, a new beach with its new barriers settled and came to bury the previous swampy deposits.

This new landscape suggests inter-dunes or inter-barriers depressions, which are seasonally or permanently flooded by the freshwaters. Once more, the pressure of oceanic water participated in the detention of the freshwater table. There are small isolated lacustrine ponds where, for several centuries, forest bushes proliferate without mangrove implements. Downstream of the outlet of Diosso amphitheatre, a small swampy forest developed on the erosion products of this amphitheatre, the rests of shrubs were dated between 600 and 500 years BP. Finally, these more or less organic deposits are covered by slightly coarse alluvial sequences, which indicate an outbreak of the erosion linked to precipitation. On this basis, but also based on various evidences such as alluvial trend at the end of the Kouilou filling, progressive reforestation of the included savannas and decrease of heliophilous components towards 500 years BP, burying of palaeo-soils near 1500 years BP, renewal of erosion of the "Série des Cirques" towards 500–600 years BP (Sitou *et al.*, 1996). Due to recent observations on Loango Bay, we can envisage a climatic evolution that is valid for a large part of this Congolese coastal margin. A similar evolution is also recorded at larger distance such as in the region of Niari, where after a flooding of the Lake Sinnda around 1300 years BP, a forested fringe settled around this lake from 650 years BP; on Plateaux

Batéké at Ngamakala, where swampy forests reappeared gradually from 900 years BP; or farther north until the Cameroon coast, where the dense Biafrean forest towards 700 years BP gradually settled in surroundings of Lake Ossa (Raynaud-Farrera *et al.*, 1996; Vincens *et al.*, 2000).

ACKNOWLEDGMENTS

We would like to thank Andrew Cooper (Ulster University, United Kingdom) and Kate Strachan (Wits University, South Africa) for valuable assistance and for constructive comments.

REFERENCES

Anthony, E., 1983, *Holocene geomorphic evolution of the coast of southern Sierra Leone*. Thèse 3ᵉ cycle, Univ. Strasbourg. p. 169.

Bertaux, J., Schwartz, D., Vincens, A., Sifeddine, A., Elenga, H., Mansour, M., Mariotti, A., Fournier, M., Martin, L., Wirmann, D. and Servant, M., 2000, Enregistrement de la phase sèche de l'Afrique centrale vers 3000 ans BP par la spectrométrie IR dans les lacs Sinnda et Kitina (Sud-Congo). In *Dynamique à long terme des écosystèmes forestiers intertropicaux*, edited by Servant, M. and Servant-Vildary, S., (Paris: UNESCO) pp. 43–49.

Bostoen, K., Clist, B., Doumenge, C., Grollemund, R., Hombert, J.-M., Koni Mulawa, J. and Maley, J., 2015, Middle to Late Holocene Paleoclimatic Change and the Early Bantu Expansion in the Rain Forests of Western Central Africa. *Current Anthropology.* **56**(3), pp. 354–383.

Chappell, J. and Shackleton, N.J., 1986, Oxygen isotopes and sea level. *Nature*, **324**, pp. 137–140.

Dechamps, R., Guillet, B. and Schwartz, D., 1988, Découverte d'une flore forestière mi-Holocène (5800–3100 BP) conservée in situ sur le littoral ponténégrin (R.P. du Congo). *C. R. Académie des Sciences,* Paris, **306**, II, pp. 615–618.

Delibrias, G., Giresse, P. and Kouyoumontzakis, G., 1973, Géochronologie des divers stades de la transgression holocène au large du Congo. *C.R. Académie des Sciences,* Paris, **276**, pp. 1389–1391.

Elenga, H., Schwartz, D. and Vincens, A., 1992, Changements climatiques et action anthropique sur le littoral Congolais au cours de l'Holocène. *Bulletin Société Géologique de France*, **163**, pp. 85–90.

Elenga, H., Schwartz, D., Vincens, A., Bertaux, J., de Namur, C., Wirrmann, D. and Servant, M., 1996, Diagramme pollinique du Lac Kitina (Congo): mise en évidence de changements paléobotaniques et paléoclimatiques dans le massif forestier. *C.R. Académie des Sciences*, Paris, **323**, IIa, pp. 403–410.

Elenga, H., Vincens, A., Schwartz, D., Fabing, D., Bertaux, J., Wirrmann, D., Martin, L. and Servant, M., 2001, Le marais estuarien de la Songololo (Sud Congo) à l'Holocène moyen et récent. *Bulletin Société Géologique de France,* **172**(3), pp. 359–366.

Fredoux, A., 1980, Étude palynologique de quelques sédiments du Quaternaire ivoirien. In Recherche française sur le Quaternaire, INQUA, *Bulletin Association Française Etudes du Quaternaire*, **1**(5), pp. 181–186.

Giresse, P., 1981, Les sédimentogenèses et les mophogenèses quaternaires du plateau et de la côte du Congo en fonction du cadre structural. *Bulletin de l'Institut Fondamental d'Afrique Noire*, **43**, A(1–2), pp. 43–68.

Giresse, P. and Kouyoumontzakis, G., 1971, Géologie du sous-sol de Pointe-Noire et des fonds sous-marins voisins. *Annales Université de Brazzaville*, **7**, pp. 95–107.

Giresse, P. and Kouyoumontzakis, G., 1974, Observations sur le Quaternaire côtier et sous-marin et des régions limitrophes. Aspects eustatiques et climatiques. *Association Sénégalaise Études Africaines., Bull. Liaison*, Sénégal, **42–43**, pp. 45–61.

Giresse, P. and Le Ribault, L., 1980, Contribution de l'étude exoscopique des quartz à la reconstitution paléogéographique des derniers épisodes du Quaternaire littoral du Congo. *Quaternary Research*, **15**, pp. 86–100.

Giresse, P. and Moguedet, G., 1982, Chronoséquences fluvio-marines de l'Holocène de l'estuaire du Kouilou et des colmatages côtiers voisins du Congo. In Les rivages tropicaux—Mangroves d'Afrique et d'Asie, CEGET-CNRS, *Travaux et Documents de Géographie Tropicale*, **39**, pp. 21–46.

Giresse, P., Malounguila-Nganga, D. and Delibrias, G., 1984, Rythmes de la transgression et de la sédimentation holocènes sur les plates-formes sous-marines du sud du Gabon et du Congo. *C.R. Académie des Sciences,* Paris, **299**, II(7), pp. 327–330.

Imbrie, J., Hays, J.D., Martinson, D.G., McIntyre, A., Mix, A.C., Morley, J.J., Pisias, N.G., Prell, W.L. and Shackleton, N.J., 1984, The orbital theory of Pleistocene climate: Support from a revised chronology of the marine ^{18}O record. In *Milankovitch and Climate: Understanding the Response to Astronomical Forcing*, edited by Berger, A., Imbrie, J., Hays, J., Kukla, G. and Saltzman, B., (Dordrecht: D. Reidel Publishing Company), pp. 269–305.

Jamet, R. and Rieffel, J.M., 1976, Notice explicative 65. *Carte Pédologique du Congo à 1/200.000*. Feuille Pointe-Noire. Feuille Loubomo 2 cartes, ORSTOM, Paris, **65**, p. 175.

Jansen, F., Giresse, P. and Moguedet, G., 1984, Structural and sedimentary geology of the Congo and southern Gabon, a seismic and acoustic reflection survey. *Netherlands Journal of Sea Research,* **17**(2–4), pp. 364–384.

Maley, J., 1997, Middle to Late Holocene changes in tropical Africa and other continents: Paleomonsoon and sea surface temperature variations. In *Third millennium BC climate change and Old World collapse*, NATO ASI Series, Global Environmental Change, edited by Dalfes, H.N., Kukla, G. and Weiss, H., (Berlin: Springer-Verlag), pp. 611–640.

Maley J., 2012, The fragmentation of the African rain forests during the third millenium BP: palaeoenvironmental data and palaeoclimatic framework. Comparison with another previous event during the LGM. *Colloque de l'Académie des Sciences*, Paris, Coforchange. Poster 2-Holocene.

Malounguila-Nganga, D., Nguié, J. and Giresse, P., 1990, Les paléoenvironnements quaternaires du colmatage de l'estuaire du Kouilou (Congo). In *Paysages quaternaires de l'Afrique Centrale Atlantique*, edited by Lanfranchi R. and Schwartz D., (Paris: Coll. Didactique, ORSTOM), pp. 89–97.

Moguedet, G., Bongo-Passi, G., Giresse, P. and Schwartz, D., 1986, Corrélations entre sédiments quaternaires continentaux et marins au Congo. *Revue de Géologie dynamique et de Géographie physique*, **27**(2), pp. 131–140.

Raynaud-Farrera, I., Maley, J. and Wirrmann, D., 1996, Végétation et climat dans les forêts du Sud-Ouest Cameroun depuis 4770 BP: analyses polliniques des sédiments du lac Ossa. *C.R. Académie des Sciences*, Paris, **322**, IIa, pp. 749–755.

Schwartz, D., 1985, Histoire d'un paysage: le Lousséké. Paléoenvironnements quaternaires et podzolisation sur sables Batéké (quarante derniers millénaires, région de Brazzaville, R.P. Congo). Thèse Doctorat es Sciences Naturelles, Université de Nancy I, p. 211.

Schwartz, D., Mariotti, A., Trouvé, C, Van Den Borg, K. and Guillet, B., 1992, Étude des profils isotopiques ^{13}C et ^{14}C d'un sol ferrallitique sableux du littoral Congolais. Implications sur la dynamique de la matière organique et l'histoire de la végétation. *C.R. Académie Sciences*, Paris, **315**, pp. 1411–1417.

Schwartz D., Guillet, B. and Dechamps, R., 1990, Étude de deux flores forestières mi-holocène (6000–3000 BP) et subactuelle (500 BP) conservées *in situ* sur le littoral Ponténégrin (Congo). In *Paysages quaternaires de l'Afrique Centrale Atlantique*, edited by Lanfranchi, R. and Schwartz, D., (Paris: Coll. Didactique, ORSTOM), pp. 283–297.

Séranne, M., Lacan, L., Bruguier, O., Giresse, P., Zahie Anka P. and Moussavou, M., 2011, Coupled sedimentary evolution of the Congo Basin and equatorial west-African Margin: A Cenozoic record of tectonic and climate forcings. *Geophysical Research Abstracts* **13** (General Assembly European Geosciences Union, Vienna, Austria).

Shackleton, N.J., 1987, Oxygen isotopes, ice volume and sea level. *Quaternary Science Reviews*, **6**, pp. 183–190.

Sitou, L., Schwartz, D., Mietton, M. and Tchikaya, J., 1996, Histoire et dynamique actuelle des cirques d'érosion du littoral d'Afrique centrale. Une étude de cas: Les cirques du littoral ponténégrin (Congo). Symposium *Dynamique à long terme des écosystèmes forestiers intertropicaux*, CNRS-ORSTOM, pp. 187–191.

Sowunmi, M.A., 1981. Aspects of Late Quaternary vegetational changes in West Africa. *Journal of Biogeography*, **3**, pp. 457–474.

Thièblemont, D., 2012, Evidence for an aeolian origin of the Holocene lateritic surface cover of Gabon (Central Africa). *Quaternary International,* **296**, pp. 176–197.

Vincens, A., Buchet, G., Elenga, H., Fournier, M., Martin, L., de Namur, C., Schwartz, D., Servant, M. and Wirrmann, D., 1994, Changement majeur de la végétation du Lac Sinnda (vallée du Niari, Sud-Congo) consécutif à l'asséchement climatique holocène supérieur: apport de la palynologie. *C.R. Acadèmie Sciences*, Paris, **318**, II, pp. 1521–1526.

Vincens, A., Elenga, H., Reynaud-Farrera, I., Schwartz, D., Alexandre, A., Bertaux, J., Mariotti, A., Martin, L., Meunier, J.-D., Nguetsop, F., Servant, M., Servant-Vildary, S. and Wirrmann, D., 2000, Réponse des forêts aux changements du climat en Afrique Atlantique Équatoriale durant les derniers 4 000 ans et héritage sur les paysages végétaux actuels. In *Dynamique à long terme des écosystèmes forestiers intertropicaux,* edited by Servant, M. and Servant-Vildary, S., (Paris: UNESCO), pp. 381–387.

Vincens, A., Elenga, H., Schwartz, D., De Namur, C., Bertaux, J., Fournier, M. and Dechamps, R., 2000, Histoire des écosystèmes forestiers du Sud-Congo depuis 6000 ans. In *Dynamique à long terme des écosystèmes forestiers intertropicaux*, edited by Servant, M. and Servant-Vildary, S., (Paris: UNESCO), pp. 375–379.

CHAPTER 10

Geochronology and technological development: The microscopic and metric evidence from Middle Stone Age (MSA) points at Mumba rock-shelter, northern Tanzania

Pastory G.M. Bushozi
Department of Archaeology and Heritage University of Dar es Salaam, Tanzania

Luis Leque
Natural Museum, Madrid, Spain

Audax Mabulla
National Museums of Tanzania, Tanzania

ABSTRACT: The material culture of Middle Stone Age (MSA) people is believed to represent a significant step towards the development of modern human behaviour. Among the most important sites, which contain human remains and archaeological records of the MSA in Tanzania, is Mumba rock-shelter. Mumba contains a more or less continuous archaeological record spanning from the MSA to historic period. Therefore, it offers unique opportunities to test hypotheses regarding the behavioural capabilities of MSA people as reflected by lithic artefact technologies and types between about 130 ka and 70 ka. In addition, three isolated molar teeth of modern humans were found in Bed VI-B of Mumba rock-shelter sequence dating to about 130 ka. This suggests that early African modern humans existed in this region during the Last Interglacial Maximum when most of tropical Africa experienced short episodes of intense rainy seasons and prolonged series of dry as well as arid environment. This suggestion of human survival in an unfriendly environment is believed to influence the developments of lithic technology, in particular, the invention of projectile technologies in Sub-Saharan Africa around 100 ka. This paper presents preliminary results on geochronology of Mumba rock-shelter and microscopic studies of Mumba MSA points. Microscopic results and TCSA (Tip Cross-Section Area) value show that MSA points were likely used as hunting devices, but sometimes they were curated, retooled and used to perform other activities. A Kolmogorov-Smirnov (KS) test indicated a gradual and trivial change over time of MSA points, but more multidimensional studies are encouraged before making a meaningful conclusion.

10.1 BACKGROUND INFORMATION

Mumba (3° 32′18″ S, 35° 17′48″ E) is one of the four rock-shelters located along the Laghangarel-Ishimijega hills in Mang'ola Chini. It is located about 1050 m a.s.l. on the eastern side of Lake Eyasi in northern Tanzania, approximately 62 km south of Olduvai

Gorge (Figure 1). The Lake Eyasi basin is one of the oldest branches of the Gregory Rift Valley in northern Tanzania where six specimens of archaic *Homo sapiens* and three isolated molar teeth of early modern humans were found (Margit and Kohl-Hansen, 1943; Pickering, 1961; Baker *et al.*, 1972; Bräuer and Mehlman, 1988; Mehlman, 1989; Bräuer and Mabulla, 1996; Dominguez-Rodrigo *et al.*, 2008). Such findings have contributed significantly to our current understanding of the origin and development of modern human populations.

The Lake Eyasi basin extends from the Ngorongoro highlands in the northwest to the Wembere-Manonga escarpments in the southwest. It is a typical half-graben showing a high fault escarpment on the northwest reaching about 2000 m altitude and a plain in the bottom at 1020 m (Figure 1). Crystalline Proterozoic and Archaean basement rock (gneiss, schist, granite) outcrops characterize most of the surrounding landscape, but the northern portion is partially covered by a fault escarpment (Ebinger *et al.*, 1997). Fossiliferous sediments that characterize the landscape of Mumba and lakeshore are relatively younger than the fault escarpment dating to about one million years ago (Pickering, 1961; Ebinger *et al.*, 1997; Foster *et al.*, 1997). However, sedimentary deposits of Mumba are different compared to the one exposed across the lakeshore. At the lakeshore, there are three sedimentary units: red soils, white sands, and green clays (Mehlman, 1989). The red sedimentary unit is famous for a wide representation of fauna and human remains (Kohl-Larsen, 1943; Rafalski *et al.*, 1978; Protsch, 1981; Mehlman, 1989; Mabulla, 1996). It is overlain by white sands and green clays units. The green clay unit contain *in-situ* and reworked animal fossil remains, lithic and a frontal bone of archaic human skull (Dominguez-Rodrigo *et al.*, 2008).

Mumba rock-shelter is part of the diorite and gneiss horst that outcrop at sides of Lake Eyasi block. A big gneissic block reaching more than 10 m high creates a protected area of around 300 m² where sediments have been accumulating for thousands of years (Mehlman, 1989; Prendergast *et al.*, 2007). The composition of alluvial, lacustrine and aeolian sediments in Mumba rock-shelter has been correlated to the existence of a previous fan-delta through which proto-Barai River out flowed (Rafalski *et al.*, 1978; Dominguez-Rodrigo *et al.*, 2007). The rock-shelter is located

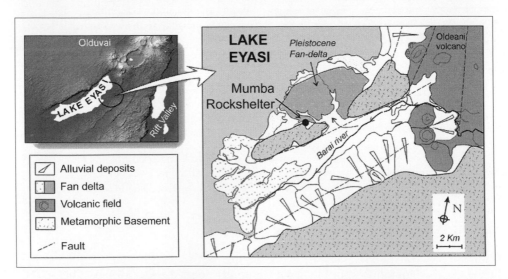

Figure 1. Location of Mumba rock-shelter and Pleistocene alluvial deposits that characterize the landscape of the Lake Eyasi Basin (from Dominguez-Rodrigo *et al.*, 2008).

at the middle of the Laghangarel-Ishimijega hills, which run parallel to the dominant northeast-southwest tectonic direction and separate the lakeshore from the current course of Barai River (Figure 1). In the rainy seasons, the Lake Eyasi extends close to the shelter catchments, but in dry seasons, the shore line retreats to about 4 km away.

Volcanic activities were intense in this part of the Rift Valley for the most of Early and Middle Pleistocene (Ebinger *et al.*, 1997). Sedimentary deposits along the lakeshore and at Mumba suggest that volcanism was remarkably reduced during the last phases of the Middle and Upper Pleistocene. The only remaining signs of volcanism are a couple of highly weathered tephra layers that are exposed in the fossil bearing sediments in the northern part of the lakeshore. Sedimentary exposures indicate fluctuations in the lake levels over time, but dynamic and systematic reconstructions of a high lake level stand for Lake Eyasi and their implications on the Pleistocene environments are lacking. However, sedimentary deposits suggest for the existence of short-term cyclic episodes that could be correlated to the effect of climatic events like *El Niño* or short intense rainfalls that may have resulted to over-flooding. In 10^4 years' scale, lake level changes can be interpreted from sedimentary changes, edaphic features, geochemistry changes, and presence of stromatolites (Dominguez-Rodrigo *et al.*, 2007).

For nearly seven decades, research at Mumba has contributed to our current understanding of the development of behaviour in early modern humans and it is considered to be one of the most significant Late Pleistocene sites in Sub-Saharan Africa. This is indicated by the number of scientific publications dealing with human subsistence, adaptation, technology successions, and symbolic representation (Kohl-Larsen, 1958; Mehlman, 1989; Conard and Marks, 2006; Prendergast *et al.*, 2007; Diez-Martin *et al.*, 2009; Gliganic *et al.*, 2012). There have been four significant excavations at the site. The first was conducted by the German pioneers between 1934 and 1938. In total they excavated about 1000 cubic metres of deposits, exposed about 9 m depth of sediments to the bottom of the site and collected a large number of artefacts (Conard, 2012). However, their collection and sorting procedures were biased towards big and well-finished lithic tools and disregarded the microlithic tools and debitage (Mehlman, 1989; Prendergast *et al.*, 2007). The density and size of discarded lithic artefacts close to the shelter clearly indicate that Kohl-Larsen's (1958) sorting significantly affected previous interpretations of Mumba archaeological sequence. Despite this fact, Kohl-Larsen's collections and their works have caught the attention of contemporary archaeologists.

The second set of excavation at Mumba was conducted by Michael Mehlman in the late 1970s and early 1980s. His excavation aimed at recovering data to address the biases in Kohl-Larsen collections and their interpretations (Mehlman, 1989). Circumstances prevented him from analysing the data he collected. As a substitute, his PhD dissertation relied much on museum collections collected by Kohl-Larsen in the 1930s, which are stored in various Museums in Germany.

The third excavation was conducted in 2005 by a team of Tanzanian and Spanish scientists. According to Prendergast *et al.* (2007), four trenches were opened around the perimeter of the major excavation area, in the most undisturbed part. However, lower sequences were not exposed because they were not part of the research agendas (Figure 2).

The fourth excavation was conducted in March 2014 by the authors. The excavation was aimed at exposing MSA sequences in Beds VI-A and VI-B. We opened up a trench in the middle of units 5 and 7 around the perimeter of the major excavation area (Figure 2). We used Mehlman's (1989) excavation profile, which is still visible as

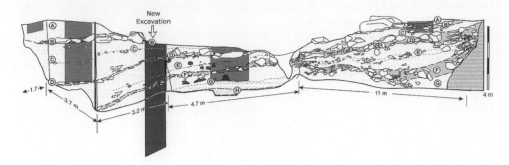

Figure 2. Stratigraphic sketch showing Prendergast *et al.* (2007) archaeological trenches (dark shaded) and the 2014 trench (light grey).

well as the site plan suggested by Prendergast *et al.* (2007) to place our trench where we believed it was the most undisturbed area. As noted by Prendergast and co-workers (2007), the geological orientation varied from previous interpretations due to sloping topography. However, we were able to cover a complete vertical sequence from Beds I to upper unit of Bed VIB. Excavation procedures followed both cultural and natural strata whereby cultural sequences were subdivided into 10 cm spits. Big bones, arte-facts and their associated height of geological strata were measured across the trench using a total station. Excavated sediments were sieved through 5 mm sieve. All datable organic materials were also plotted using the total station. Two Achatina shells and five teeth specimens were collected from Beds V and V-A for ESR dating.

In this paper, we present and re-interpret the stratigraphic sequence of lower sequences of Mumba rock-shelter, which were not discussed in the most recent studies (Prendergast *et al.*, 2007; Diez-Martin *et al.*, 2009; Gliganic *et al.*, 2012). We also dis-cuss the trend of technological change and tool functions based on microscopic and Tip Cross-Section Area (TCSA) values on MSA points from different geo-chronological sequences of Mumba (Mehlman, 1989). We believe that comparative technological and functional analysis using one independent variable could contribute to better inter-pretations of technological and typological dynamics that have received attention by scholars (Mehlman, 1989; Conard and Marks, 2006; Diez-Martin *et al.*, 2009).

10.2 METHODOLOGICAL APPROACHES

Points were analysed by using microscopic analysis to identify possible obstructions and signals that could be used to distinguish projectile and non-projectile actions and, to some extent, material worked. Microscopic variables were delineated based on the state of tool edge morphologies, presence or absence of traces, position of traces, edge damages, striations, and polish distributions on each tool. Two electron microscopes were used to trace use-wear and impact damage patterns. An Olympus BX51 stereomi-croscope with 5x, 10x, 20x, and 50x magnification was used to trace polish, striation, and rounding patterns. An Olympus DP71 binocular microscope with 0x, 7x, and 9x magnifications was used to trace micro-scars and possible residue remains on tool edges. A magnification range of 100 and 500 diameters, using a reflected-light and differential-interferences microscope with polarized light, was used to register some use-wear signals on tool edges. At magnification greater than 100 diameters, it was possible to distinguish use-wear signals from natural edge damage or any other modifications related with flaking procedures.

To evaluate wear traces, the full ranges of magnifications were employed, from 100 to 200, and then to 500 mm. Both bright and dark field illuminations were used. A tool surface scan of both ventral and dorsal surfaces (as well as of edges of tools) was done at magnifications of 100 and even 200 mm. Further examinations at 500 mm were done whenever possible to better characterize wear traces. Orientation and location of photomicrographs were recorded relative to those of artefacts. Photomicrography size was determined by photographing a micrometre disk scale at an appropriate magnification. Wear traces were further described, even when not photographed. Location of wear traces discovered at intermediate-range magnifications was also noted in comparison between microscopic and lower-powered magnifications (Keeley, 1982). Results from microscopic studies were backed with evidence from metric data on MSA points, in particular, Tip Cross-Section Area (TCSA), and a Kolmogorov-Smirnov (KS) to explore functional and technological difference between lithic points from Beds VI-B, VI-A and V of Mumba. The TCSA is useful for identification of an impact a stone point can make after hitting a target or to approximate the size of wound that can be created by a stone point (Hughes, 1998; Shea, 2006; Shea and Sisk, 2010; Sisk and Shea, 2011). Even though some researchers prefer the use of the Tip Cross-Section Parameter (TCSP), both variables (TCSA and TCSP), have been used and have produced closely related results indicating that MSA points were used as inserts for hunting weapons (Sisk and Shea, 2011). In a KS test, variables with comparable characteristics follow the same cumulative distributions and their computed p-value is always greater than significance alpha level (Bushozi, 2011; Wilkins *et al.*, 2012).

10.3 GEOCHRONOLOGY

Sedimentary deposits at Mumba rock-shelter are composed of fallen blocks, slabs, alluvial sand, lacustrine sediments, and aeolian deposits. After a number of previous archaeological excavations, the main body of the sediments disappeared, remaining only a half ring of peripheral deposits. Commonly, such peripheral areas of the shelter infill show higher slopes, smaller block sizes and some of the layers are mainly of aeolian or gravitational origin. Taking into account that Mehlman's 1989 stratigraphic sequences were primarily developed based on the central sedimentary records, our experiences from peripheral zones differ from Mehlman's interpretations.

In previous studies, Kohl-Larsen (1943), Rafalski *et al.* (1978), and Mehlman (1989) correlated the geological strata of Mumba with lakebeds. They described six geological strata, namely, Beds I–VI, over 10.75 m depth whereby the lower Bed was subdivided into two sub-strata, Beds VI-A and VI-B. Mehlman (1989) attempted to correlate the shelter and lake deposits based on progression of archaeological materials and arbitrary layers as a reference. This differs from our current attempt, whereby geological interpretation is mainly defined based on sedimentary deposits and natural geological strata. For that reason, we argue that Mumba Beds are not horizontally analogous. They tend to fluctuate in thickness; but in any case, Mehlman's synthesis should be recognized as the most complete series obtained at Mumba. His study provided a more detailed definition of the culture-chronology of Mumba. However, in many cases, Mehlman's interpretations were highly influenced by Kohl-Larsen's stratigraphic sequences. The major contributions made by Mehlman include detailed sedimentary descriptions, reconstruction of culture-history, and dating of the beds that he defined (Table 1).

Table 1. Stratigraphic and cultural sequences and radiometric dates of Mumba.

Bed	Content description	Material	Method	Uncalibrated dates BP	Lab. No	Reference
II-I(1)	Trench 8. Iron age materials, near surface	Charcoal	AMS C^{14}	381 ± 91*	AA-69910	Prendergast et al., 2007
II-I(1)	Trench 5. Ceramic, near surface	Charcoal	ASM C^{14}	398 ± 86*	OS-61330	Prendergast et al., 2007
III-I(1)	Trench 6. Twisted-cord roulette sherds	Charcoal	ASM C^{14}	844 ± 78*	OS-61329	Prendergast et al., 2007
II-lower	LSA/Pastoral Neolithic/Iron Age	Charcoal	C^{14}	1780 ± 80	ISGS-565	Mehlman, 1989
III-2(3)	Kansyore and other sherds	Charcoal	ASM C^{14}	1769 ± 153*	AA-69911	Prendergast et al., 2007
III	Burial IX (human remains)	Charcoal	C^{14}	4860 ± 100	UCLA-1913	Mehlman, 1989
III	Burial IX (human remains)	Bone collagen	C^{14}	4890 ± 70	FRA-1	Mehlman 1989
III-upper	LSA	Quartz-grain	OSL	12,000 ± 1700	MR2	Gilganic et al., 2012
III-mid	LSA	Quartz-grain	OSL	15,600 ± 1200	MR3	Gilganic et al., 2012
III-lower	Nasera Industry	Quartz-grain	OSL	36,800 ± 3400	MR4	Gilganic et al., 2012
111-lower	Nasera Industry	Ostrich eggshell	C^{14}	26,960 ± 760	ISGS-566	Mehlman, 1989
III-lower?	Not published	Ostrich eggshell	ASM C^{14}	33,000–29,000*	Not published	Conard, 2005
IV	Beach deposits	Tufa	C^{14}	25,130 ± 320	USGS-1505	Mehlman, 1989

Stratum	Industry	Material	Method	Age	Lab no.	Reference
IV/V	Reworked Bed I	Achitina shell	C^{14}	36,900 ± 800	ISGS-499	Mehlman, 1989
V	Not published	Ostrich eggshell	AAR	33,460 ± 900	AA-3299	Brooks et al., 1990
V-upper	Mumba Industry	Quartz-grain	OSL	49,100 ± 4300	MR6	Gliganic et al., 2012
V-upper	Mumba Industry	Bone apatite	Th-230	23,680 + 1091, −851	USGS-83-10	Mehlman, 1989
			Pa-231	23,800 + 2538, −1414		
V-mid	Mumba Industry	Quartz-grain	OSL	51,300 ± 4200	MR7	Gliganic et al., 2012
V-mid	Mumba Industry	Achitina shell	C^{14}	>37,000	GX-6620A	Mehlman, 1989
V-mid	Mumba Industry	Achitina shell	C^{14}	31,070 ± 500	ISGS-498	Mehlman, 1989
V-mid	Mumba Industry	Bone apatite	Th-230	65,686 + 6049, −5426	USGS-82-22	Mehlman, 1989
V-mid	Mumba Industry	Bone apatite	C^{14}	29,570 + 1400, −1100	GX-6621A	Mehlman, 1989
V-mid	Mumba Industry	Bone apatite	Th-230	46,600 + 2050, −1725	USGS-82-33	Mehlman, 1989
V-mid	Mumba Industry	Quartz grain	OSL	39,000 ± 3000	GL14014	Provisional
V-lower	Mumba Industry	Bone apatite	C^{14}	20,995 ± 680	GX-6622A	Mehlman, 1989
V-lower	Mumba Industry	Bone apatite	Th-230	35,291 + 749, −476	USGS-82-20	Mehlman, 1989
			Pa-231	39,777 + 4162, −3753		
V-lower	Mumba Industry	Quartz-grain	OSL	56,900 ± 4800	MR7	Gliganic et al., 2012
V-lower	Mumba Industry	Quartz-grain	OSL	54,000 ± 4000	GL14013	Provisional
VI-A	Kisele Industry	Quartz-grain	OSL	63,400 ± 5700	MR9	Gliganic et al., 2012
VI-A	Kisele Industry	Quartz-grain	OSL	73,600 ± 3800	MR10	Gliganic et al., 2012
VI-A-upper	Kisele Industry	Quartz grain	OSL	71,000 ± 22,000	GL14010	Provisional
VI-B-top	Kisele Industry/MSA	Bone apatite	C^{14}	19,820 ± 750	GX-6623A	Mehlman, 1989
VI-B	Sanzako Industry/MSA/human remains	Bone apatite	Th-230	131,710 + 6924, −6026	USGS-82-19	Mehlman, 1989
			Pa-231	109,486 + 44,404, −23,020		

*Calibrated using IntCal 04 curve (Reimer *et al*., 2004).

A series of successive alternation of blocks of gravitational origin and fine aeolian sediments were revealed in this study supporting an idea raised by Prendergast *et al.* (2007) that sedimentary deposits of Mumba are characterized by fallen blocks as well as alluvial and aeolian deposits. Even though changes in sedimentary deposits can be easily used to define the stratigraphic sequence of Mumba, our interpretation should not be used as criteria to undermine previous versions. This is due to the complex nature of sediment distribution across the vicinity of the shelter. To avoid unnecessary confusions, the stratigraphic sequences proposed by Prendergast *et al.* (2007) were updated (Figure 3 and 4). Prendergast *et al.* (2007) started with Level A and ended with Level I subdivisions, which correspond to Mehlman's Bed V1-B (Figure 3). The subdivisions are helpful not only because of their clarity in test excavations, but also because Kohl-Larsen (1943) and Mehlman's (1989) stratigraphic sequences are difficult to recognize in peripheral zones.

In this study, the sedimentological and geo-chronological descriptions are presented in terms of Beds and start with Upper Bed III because the upper most Beds I and II were missing in the excavated trench (Figure 3). Archaeologically, Upper Bed III is composed of a Ceramic Later Stone Age (LSA) and it was radiocarbon dated to about 1843 ± 60 BP, but lower Bed III is composed of Aceramic microlithic LSA dating to about 36,800 ± 3400 BP (Table 1). The maximum thickness of Bed III measures to about 1.8 m. It consists of very fine sands and silts, divided in at least three sub-levels, sandy in the upper unit gradually turning to finer and more cemented, rich in gastropod (*Achatina*) shells and fauna remains including lithic artefacts in the mid portion. It slopes gently to the west and is characterized by small blocks as well as slabs. The lower unit is composed of a small strip of alluvial deposits interpreted by Kohl-Larsen and Mehlman as Bed IV (Prendergast *et al.*, 2007). Thus, the strip of alluvial deposits that separates Bed III is situated between metamorphic blocks with coated calcium carbonate of algae origin and small rounded pebbles suggesting an episode of a high lake level (Mehlman, 1989).

Generally, deposition of Beds III and IV revealed interesting connotations about weathering conditions and sedimentation processes. In some sections, in particular, in the West and Northwest sections, there is a great slope in the shelter sequence that links Beds III and IV. Practically, such an incidence can be interpreted as a cleft in sedimentation deposits resulting from the deposition of fallen blocks in Bed IV. Alternatively, it may indicate a dramatic environmental change at this age. The eastern side of the shelter show much more blocks than the western side, which is highly characterized by fine-grained sediments. There is a high slope that can be correlated to the coned-shape of the shelter's deposits sloping towards the western margins. To a great extent,

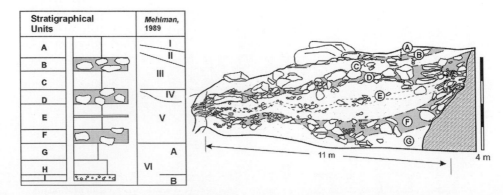

Figure 3. Stratigraphic sequences of Mumba adopted from Prendergast *et al.* (2007).

the observed rock boulders may have resulted from the fallen shelter's roof. Alluvial deposits may suggest for an increase in lake levels or accumulation of rainy water inside the rock-shelter. However, presence of bioturbation features such as hideaways, dens and burrows as well as archaeological remains suggest that water infill sustained briefly to the extent that alternatively, it may indicate presence of post-depositional processes (Figure 3). Available radiocarbon dates indicate that Bed IV has an age of 25,130 ± 320 BP (Table 1).

Upper and Middle Bed V units are about 1.2 m thick and it is a very homogeneous bed of sandy, silt, and loamy sediments (Figure 4). It includes some scarce slabs and fallen blocks, all of them dipping in the same direction. At the top of the unit, there are a number of fallen slabs and blocks. It is rich in archaeological composition and *Achatina* shells. Available radiometric dates place the upper and middle units of Mumba Beds V between 23,800 ± 2538 BP and 46,000 ± 2050 BP (Table 1). Lower Bed V is laterally, divided into two different units, indicating two different stages of formation. It is composed of fine-grained sediments with limited fallen blocks. Radiometric dates place the lower unit of Bed V between 54,000 ± 4000 and 65,689 ± 604 BP (Table 1). Lithic artefacts are characterized by flakes and retouched tools made by *Levallois*, disc, radial and bipolar core technologies, but bipolar cores predominate (Mehlman, 1989; Diez-Martin

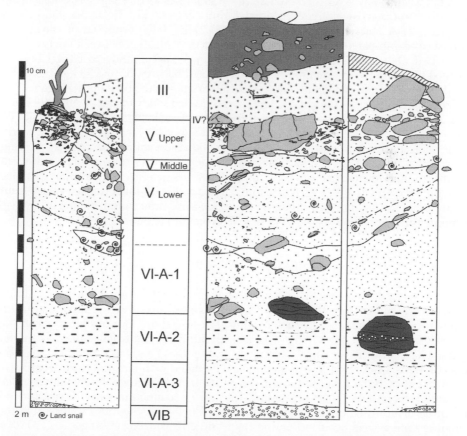

Figure 4. Stratigraphic sequence of Test Pit 8 at Mumba: Left is the southern profile; in the middle, the western; and to the right, the northern profile. Geological levels are correlated with previous interpretations. Dark shaded surfaces are signs of big burrows.

et al., 2009). Other lithic artefacts include blades; notched scrapers; side and end scrapers; sundry scrapers; large backed pieces; and bifacial, unifacial and *Levallois* points (Ambrose, 2001; Bretzke *et al.*, 2006; Mehlman, 1989). Lithic artefacts were found associated with symbolic revealing artefacts including ostrich egg shells, beads, and ochre.

The Mumba Industry has become a focus of recent debates about its cultural affiliation and the timing for evolution of cognitive abilities in human evolutionary history (Mehlman, 1989; Marks and Conard, 2006; Diez-Martin *et al.*, 2008). Mehlman (1989) defined the Mumba Industry as a transitional industry between the MSA and LSA. Conard and Marks (2006) suggested that the Mumba Industry represented a late MSA, similar to the Howiesons Poort Industry of Klasies River in South Africa. However, this idea was recently challenged by Diez-Martin *et al.* (2009) who placed the Mumba Industry to an early LSA. Early LSA Industries have also been recorded reported from the Naisuisui Beds at Olduvai Gorge, Tanzania and Lukenya Hills in Kenya (Ambrose, 2002; Skinner *et al.*, 2003; Diez-Martin *et al.*, 2009).

In this study, Bed VI-A was subdivided into three sub-units: Beds VI-A1, VI-A2 and VI-A3 (Figure 5). Bed VI-A1 is 0.8 to 1.0 m thick, and consists of poorly sorted sand and silt. Sediments are more compact but less homogeneous than the overlying Beds. It shows coarser grained sediments in the middle section, perhaps because of bioturbation. Such sandy layer is characterized by small blocks dipping horizontally, suggesting for the occurrence of a steep slope during deposition of the upper unit of Bed VI-A. Bed VI-A2 measures about 0.35 to 0.6 m thick. It is composed of clayey and compact sand compared to the overlying unit, but the transition between the two units is spiny. Bed VI-A3 is 0.55 m thick, composed of poorly sorted sand and silt soil, but not cemented like in the overlying unit. Bed VI-A is dated between 63 and 73 ka (Table 1).

Bed VI-B was exposed in the western corner extending to about 0.3 m (Figure 4). It is a thick layer of gravels characterized by coarse-grained sandy matrix of alluvial origin. A variety of bed thickness and variable distribution of sediments show the existence of palaeo-surfaces, but stratigraphic sequences do not support Mehlman's arbitrary subdivisions or archaeological pits. Real isochrones subdivision should be arched following orientation of natural stratigraphic strata. Arched depositional strata and signals resulted from macro- and micro-organism activities are most frequent in cave or shelter deposits. The middle sequence of Bed VI-B was found associated with three isolated molar teeth of early anatomically modern humans [*Homo sapiens* (Bräuer and Mehlman, 1988)]. A uranium-series on bone apatite from Bed VI-B dates to about 131 ka (Table 1).

10.4 PALAEOENVIRONMENTAL RECONSTRUCTION

Preliminary results suggest at least four changes in depositional environments of Mumba rock-shelter from Beds III to VI-B. The lowermost gravel sediment indicates a high-energy environment like fluvial, alluvial or lake shore. It looks highly probable that it was related with the activity of an ancient fan-delta that reached the rock shelter when ancient Barai River out-flowed close to Mumba. Sandy and more clayey sediments of Beds VI-A3 and VI-A2 could also be related with late stages of progressive displacement of Barai River to the current position. Such changes might have resulted from different natural factors including volcanism, sedimentological displacements, and tilting or uplifting of the landscape (Figure 1). Tectonic features suggest that the current graben system occurred during the Middle Pleistocene (Ebinger *et al.*, 1997). Another possibility is sedimentary filling of the outflow area and diversion to the lowest area in the south. Such succession of sedimentary deposits was followed by partial erosion of clayey sands that characterize Bed VI-A2 (Figure 4).

This was followed by partial erosion of Bed VI-A2 and it is probably related with the beginning of aeolian deposition in Bed VI-A1. Low content of clay deposits may also suggest for the existence of flooding at a certain point. Such observation is also supported by alternation of beds with high frequencies of blocks and slabs, and those with fine-grained sediments indicating fluctuations in climate condition at the beginning of the Upper Pleistocene. The high concentration of blocks and slabs in Beds VI-A1 could be related with periods of arid, dry, and windy conditions with short episodes of rainfall. Such climatic conditions favour evaporation of ample amount of ground moistures, soil erosion, and displacement or falling of big rocks. Also, very fine grained silty sediments that characterize the upper part of Bed VI-A1 could be related with dry periods when soil deflation occurs and aeolian deposition dominates. The upper units of Beds VI-A1 indicate the development of a steep slope widely exposed in the western side of the shelter. This is followed by the loamy surface that characterized lower Bed V. At that stage, the angle of slope increased up to 50°–60°, suggesting for an episode of dry season. Again, it is followed by periods of humid climatic conditions characterized by big fallen blocks coated with algae between 25 and 35 ka correlated to Bed IV (Mehlman, 1989). In this case, the lake level increased reaching the rock shelter and causing the deposition of alluvial sands and rolling stones interpreted by Mehlman (1989) as a beach deposit. Bed III is composed of fine grained aeolian deposits that may have been influenced by changes in wind direction, which resulted in accumulation of sediments in south-western direction.

Previous studies on Middle Pleistocene deposits along the lakeshore indicate dominance of terrestrial pollen excluding the herbaceous aquatic plants (Dominguez-Rodrigo *et al.*, 2007). Such pollen was from tuffaceous clayey thin layer and fan delta sediments widely distributed in the ancient-Barai River alluvial plain and delta. As it was noted in previous sections, there are three fossiliferous units—red soils, superimposed by white sands, and green clays (Dominguez-Rodrigo *et al.*, 2008). White sands and green clayey deposits were recently ESR-dated to about 118 + 14 ka, uranium series gave the age estimate of about 125 ± 3 ka, suggesting that it was deposited during the Last Interglacial Maximum (Dominguez-Rodrigo *et al.*, 2008). Dates for red fossiliferous unit famous for wide representation of fauna and human remains is unknown but Manega (1994) correlated it with the Ndutu (Olduvai Gorge) and Ngaloba Beds (Laetoli) that are approximated to date between 200 ka and 250 ka. Recovered fauna fossil remains including bones of hippopotamus, lions, hyena, bovid or equids, ostrich, crocodiles, and cat fish (Dominguez-Rodrigo *et al.*, 2008) overall water dependant species and genera suggest for existence of freshwater environment. Lake Eyasi was not saline like it is at the present (Viences *et al.*, 2005).

Pollen samples from the upper fossiliferous units yielded about twenty-one distinct pollen taxa and two non-pollen species (Dominguez-Rodrigo *et al.*, 2008). Herbaceous taxa predominate (61.4% to 67.5%); among them Poaceae is dominant (46.6% to 52.6%). Other important herbaceous taxa identified include Chenopodiaceae, Amaranthaceae, and Tubuliflorae (3.1% to 4.6%) as well as Plantago (0.6% to 1.1%) (Dominguez-Rodrigo *et al.*, 2007). Among the modern arboreal member of the Somalia-Masai phytogeographic region (10.8% to 14.5%), three pollen types were identified whereby Acacia (1.6% to 2.8%), *Commiphora africana* (0.9% to 10.6%) and *Salvadora persica* (6.9 to 10.6%). Comparing these data with modern environment (Vincens, 1987), suggests that in the past, *Salvadora persica* species were commonly found within a more arboreal range of the surrounding environment than it is found at the present (Dominguez-Rodrigo *et al.*, 2007). This does not mean that *Salvadora* was dominant over *Acacia* or *Commiphora*, since both are weaker pollinisers, while *Salvadora* is a stronger polliniser.

Generally, the pollen types suggest that for the most of Upper Pleistocene, the local environment in the Eyasi basin was very open, dominated by grasses. Woodlands and shrubs were few and occurred only in riverine catchments (Dominguez-Rodrigo *et al.*, 2007). Thus, vegetation cover in Lake Eyasi basin and surrounding landscape was very similar to one that can be observed in the area today, that is, extended open grasslands with scattered shrubs, which are elements of Somalia-Masai phytogeographic region, narrow arboreal riparian formation along the river or lake catchments, and finally, Afromontane forest in highlands and mountains surrounding the lake basin (Vincens, 1987). If recent dates from upper taffaceous deposits (white sands and green clayey) sediments are correct, pollen samples that have been represented could belong to the Eemian interglacial (OIS 5e) between 115 and 130 ka. These dates also correlate to the lower sequences of Mumba, in particular, Beds VI-B (Table 1). The high proportion of grasses equally could be interpreted as the result of an arid-phase whereby lake levels decreased and areas that were formally submerged were exposed because of descended lake catchments. However, reliable chronological reconstruction of Mumba and Lake Eyasi basin are urgently needed to make credible inferences about past environmental settings and its influence on human cultural and technological developments.

Though climate was not the primary contributory factor for technological innovations, there is little doubt that it has played a significant role in the history of human evolution, particularly, in early stages, when all humanity depended upon gathering of plant foods and occasional hunting of wild animals. Several researchers have also argued for the deep-sealed dynamics between climatic changes and their effect on Pleistocene environment including technological development (Cohen *et al.*, 2007). It has been argued that changes in climate promoted a discernible pattern of cultural responses. For instance, core sediments from Lakes Malawi and Tanganyika in the Western Rift Valley System alongside archaeological records from Mumba and other sites in the Eastern Rift Valley System make broad correlations with distinct cultural developments occurring in both fine and chaotic times (Cohen *et al.*, 2007).

10.5 ENVIRONMENTAL DYNAMICS AND TECHNOLOGICAL CHANGE

To a great extent, deteriorated environments may cause animal populations to become severely reduced and survivors to be scrawny. In hunting and gathering communities, such situations do not outlaw the necessity of meat eating because other sources of food were scarce as they were similarly affected by climatic instability. Under such circumstances, meat remained a dependable source of protein and fat nutrients during unfriendly environmental conditions. For that reason, it is proposed that advanced hunting technologies emerged as a human response to environmental deteriorations that characterized tropical Africa for the most parts of Upper Pleistocene (Scholz *et al.*, 2007; Shea and Sisk, 2010). Possibilities for ambush, trapping, and scavenging were obvious at the beginning of drought episodes when scrawny animals were easily attained. In good seasons, survivors from mass extinction were retreated in few isolated refuges, thereby hunting opportunities were severely reduced. Therefore, new hunting strategies and weapons that can kill animals at a distance were innovated. It was during that stage when MSA points became one of potential killing tools (Hughes, 1998; Sisk and Shea, 2011). Stone points were made to be fixed at the end of organic shafts to form spearheads and later on, the arrowheads.

As noted previously, the model to test for diversification of Stone Age hunting weapons was developed by Thomas (1978) based on ethnographic hunting weapons in South America and later modified by Hughes and Shea in 1990s to test Stone Age hunting weapons globally (Table 2). East and South Africa are believed to be the ancestral homeland of early modern humans. Available lithic evidence from these regions indicates that throwing spears predominated between 125 ka and 60 ka, followed by bow and arrow in later phases (Villa and Lenoir, 2009; Lombard and Phillipson, 2010; Shea and Sisk, 2010; Sisk and Shea, 2011). Here we continue along this line of investigation, combining metric measurements and use wear evidences. To give an idea of the metric variation projectile technologies, about 104 points from Bed VI-B (28), Bed VI-A (29), and Bed V (49) were analysed. Points were further investigated through examination of use-wear and hafting traces (Lombard, 2007; Wurz and Lombard, 2007; Bushozi, 2011). Even though the sample size analysed here was very limited for statistical testing, initial results raise interesting inferences with regard to functional interpretations of archaeological tools.

Samples for microscopic studies were dominated by quartz raw materials, only two (13.3%) were made from quartzite. The majority of points were abraded or characterized with micro-breakage patterns due to deposition, post-deposition, and storage processes. About 26.6% of the analysed sample had evidence of use-wear signals related to projectile actions such as impact edge damage, bending fractures, striations resulting from longitudinal or vertical pressure in distal or proximal ends, burination, and wear polishes or oxides (Grace, 1989; Sano, 2009). However, this does not directly imply that these fracture patterns are automatically diagnostic evidence of projectile use because they could also be due to other factors (Sano, 2009). Damage related to projectile actions often occurred close to the tip end or penetration angle (Grace, 1989). Sometimes, they formed symmetrical striations, directed towards the striking edge or striations oblique to the edge, which means a possible transverse motion (Figures 5, 6 and 7). Symmetrical striations related to projected actions are like those revealed on tip, fluted or burin-like impact fractures are visible on proximal end 5 (Figure 5). Fractures related to projectile actions are evident at the distal ends of a tool microphotography 1 and 2 (Figures 6 and 7). Fluted burinations (Grace, 1988), symmetrical striations, and micro-fracture signals observed on samples from Mumba rock-shelter are considerable diagnostics for hunting related activities (Grace, 1989; Sano, 2009; Lombard and Phillipson, 2010).

Nevertheless, use-wear signals indicate that points performed other non-projectile activities. Use-wear signals that occur on points are smooth and flat

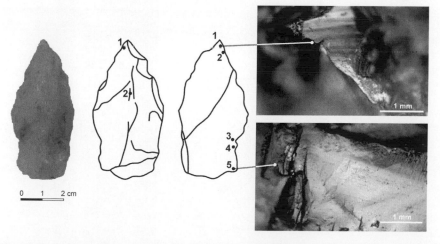

Figure 5. Impact striations on top and microscopic haft damage below on dorsal surface, viewed at 200X diameter Photographs 1 and 2 indicate impact striations that occurred on the dorsal and ventral surfaces produced by projectile actions. Photographs on top indicate fluted scar that is also indicative of a projectile action. Other fractures on proximal ends are associated with non-use processes.

polishes characterized by longitudinal motion indicating that they could have been produced by cutting actions (Keeley, 1982). Cutting actions produce longitudinal striations that are parallel to the cutting edge, and they account for at least 20 percent of the analysed sample. About 6.7% of the analysed samples have rounded

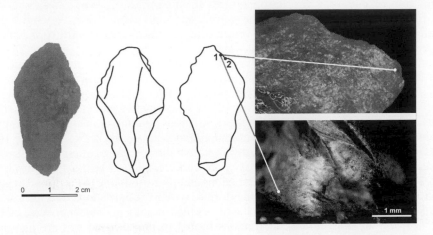

Figure 6. Microscopic photograph at 200X. Photographs 1 and 2 on the interior surface indicate a spin-off breakage around the tip, which can be correlated with impact damage. Small striations were found at the ventral side-right edge, parallel to the edge, but polishes are lacking, probably because of post-depositional processes.

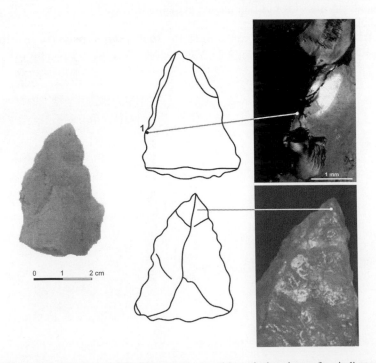

Figure 7. Microscopic fracture at 200X. Photographs 1 and 2 on the interior surface indicate a possible burination at the right edge, which can be related to impact damage.

polishes perpendicular to their working edges similar to those produced by actions on wood, suggesting sometimes stone points were used as burins (Figure 8). Points with linear marks parallel to the edge and little smooth spot or flat polishes similar to those caused by working on bone make up 6.7%. There was a fair representation of points that were severely affected by deposition and/or post-depositional processes, accounting for about 13.3% of the total. Edges of these tools are highly damaged and their surfaces are characterized by a number of chaotic striations (Figure 9). One point was found with micro-fracture that could result from accidental breakage during excavation or crushing due to improper storage (Figure 10).

As it was mentioned in previous sections, MSA points from Mumba were further analysed based on the TCSA value and KV test to perceive intended functional roles

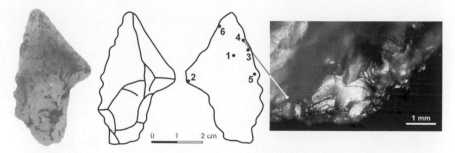

Figure 8. Microscopic photograph at 500X on the ventral with rounded polish, similar to those provoked by action on wood.

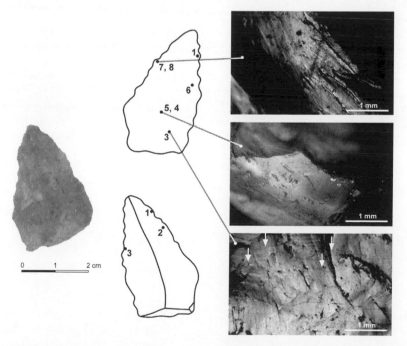

Figure 9. Microscopic photograph at 200X on the ventral dorsal surfaces indicating micro-breakage and chaotic striations resulting from deposition and post-disposition or storage processes.

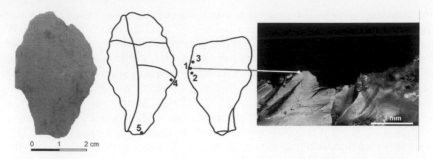

Figure 10. Microscopic photograph at 200X on the interior surface indicating very recent micro-fracture.

Table 2. Mean TCSA and SD of ethnographic, experimental, and archaeological projectiles
(ka ≈ 1000 years).

Weaponry systems	Age	Frequency	TCSA	SD	Sources
Ethnographic arrowheads	–	118	33 mm²	20	Thomas, 1978; Shea, 2006, 2009
Experimental arrowheads	–	–	47 mm²	–	Hughes, 1998
Ethnographic darts	–	40	58 mm²	20	Hughes, 1998; Shea, 2006, 2009
Ethnographic thrusting spears	–	28	168 mm²	89	Shea, 2006
Levant UP (Emireh points)	30–45 ka	47	132 mm²	85	Shea, 2006
Levant MP points	80–130 ka	749	135 mm²	67	Shea, 2006
Kebara IX-XII Levallois points	–	61	55 mm²	19	Shea, 2009
Epic Cave unifacial points	77.5 ka	76	45 mm²	9	Ambrose, 2002; Shea, 2009
Epic Cave bifacial points	77.5 ka	22	49 mm²	9	Ambrose, 2002; Shea, 2009
Cartwright's MSA points	–	74	91.1 mm²	36.7	Waweru, 2007
Rose Cottage MSA points	57–33 ka	74	78 mm²	33	Mahopi, 2007; Villa and Lenoir, 2009
Sibudu Cave MSA points	60–36 ka	21	116.2 mm²	41.5	Villa and Lenoir, 2009
Klasies River MSA 1-II points	115 ka	150	86 mm²	19	Shea, 2009
Blombos Cave Still Bay points	73–140 ka	90	45 mm²	20	Shea, 2009
Aterian tanged points	30–50 ka	29	59 mm²	13	Shea, 2009
Magubike MSA points	150 ka	117	82.7 mm²	40.9	Bushozi, 2013
Nasera MSA points	56 ka	16	80.2 mm²	35.3	Bushozi, 2011
Mumba Bed VI-B	135 ka	28	52.2 mm²	38.1	Analysed sample
Mumba Bed VI-A	109 ka	29	47.4 mm²	18.7	Analysed sample
Mumba Bed V	58 ka	47	51.4 mm²	38.7	Analysed sample

and trend of technological change over time. The mean scores TCSA for the points from Mumba were further compared with ethnographic, experimental and archive evidence across Sub-Saharan Africa to examine consistencies in manufacture and use of hunting weapons (Table 2). When compared to the ethnographic and experimental points, the sampled points from Mumba ranging from 52.2 mm^2 to 47.4 mm^2. That is

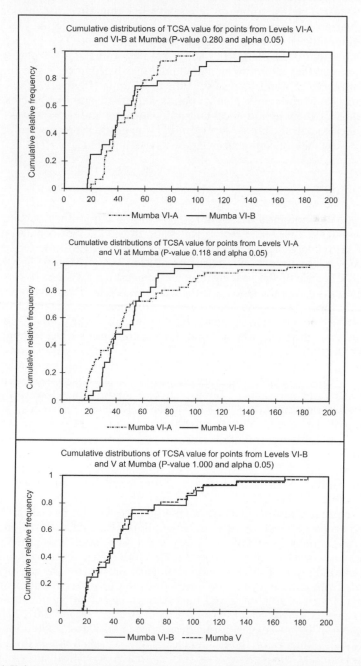

Figure 11. A Kolmogorov-Smirnov (KS) test for points from Beds VI-B and VI-A and V at Mumba.

a larger TCSA than the mean range arrowheads, which measure between 33 mm² and 47 mm². However, this is a little bit smaller than the mean range of throwing spears or darts that measures to about 58 mm² (Thomas, 1978; Hughes, 1998; Shea, 2006, 2009; Shea and Sisk, 2010).

Furthermore, discrepancies for points from the three stratigraphic sequences of Mumba (Sanzako, Kisele, and Mumba) were statistically tested using the Kolmogorov Sminorv (KS) test. The test compares differences in cumulative distributions for two independent variables. However, techno-typological variations defined by Mehlman (1989) remain unnoticeable when looking at the TCSA (Figure 11). The KS test indicates that points from Beds VI-B, VI-A and V have a *p*-value of ≤ 1.0 and an alpha of 0.05, indicating that they all follow similar cumulative distribution suggesting for continuity and similarity in point forms over time at Mumba (Figure 11). Similarities in the TCSA values at Mumba indicate that points were made and used in closely related ways, regardless of their hypothetical cultural traditions and age settings. This may suggest that technological changes went through the gradual process and it persisted for the long period, before commencement of new equipment.

10.6 DISCUSSION

Preliminary geo-chronological descriptions and cultural chronology of Mumba rockshelter is characterized by fallen blocks and slabs embedded between alternating silt-sand horizons (Prendergast *et al.*, 2007). Our sample for the ESR and OSL dating are still in process, but provisional OSL dates place Beds VI-A of Mumba between 63 and 93 ka (Table 1). Signals of the past environment suggest for existence of local environmental fluctuations across the landscape that surrounds Lake Eyasi Basin for the most of Middle and Upper Pleistocene. The role of environmental change in subsistence patterns and cultural innovations has been a subject of intense studies and debate. However, most scientists share the basic principle that climatic conditions affect resource richness, which, in turn, set the carrying capacity of an ecosystem (human population that can be supported in a given region, deMenocal and Stringer, 2016). This then guides the trend of cultural innovations including strategies to acquire basic subsistence requirements.

Sedimentary deposits suggest for existence of a river flow close to the shelter's catchment area during the early phases of Upper Pleistocene. The ancient river would have been one of the reasons that attracted the long stay of Pleistocene forager communities at Mumba. Water and other resources that depended on water, like vegetation and hunted prey, were easily found (Mehlman, 1989; Mabulla, 1996; Prendergast *et al.*, 2007; Mabulla, 2007). However, more reliable dates are needed in the lower sequences of Mumba to correlate putative environmental fluctuations with typo-technological dynamics.

Similar complications such as fragmented palaeoclimatic records and a paucity of well-dated archaeological deposits are affecting the East African region in general (McBrearty and Brooks, 2000; Gliganic *et al.*, 2012). Most sites were either dated more than a decade ago or have not been dated at all. Under such conditions, it is difficult to compare chronologies of different archaeological sites across the region. However, proxy records from sediment cores collected from the East Arc mountains in Kenya and Tanzania (Finch *et al.*, 2009), and East African lakes—in particular, Lakes Tanganyika, Malawi, Masako, Naivasha and Challa (Cohen *et al.*, 2007; Scholz *et al.*, 2007; Finch *et al.*, 2009)—suggest for inconsistence in lake level fluctuations probably because of direct influence of the Indian Ocean and biannual north-south migration on the Inter-Tropical Convergence Zone (ITCZ) that govern the distribution of moisture and rainfall in the region. In due regard, complexity of atmospheric ecosystem can be seen as a major cause of local climatic variations and vegetation dynamics that characterized the region today (Tierney *et al.*, 2011). Even though quantification of environmental dynamics and human response is a complicated phenomenon, interaction of several atmospheric systems provides some useful details in terms of what enabled East Africa to host a number of refugees during the Pleistocene climatic instability and sustaining early modern humans (Willoughby, 2007, 2012).

It has been frequently argued that different hunting weapons were used depending on the ecological setting and prey choice (Shea, 2009). Thus, spear weapons were mostly preferred in woodland or forestry, while arrows were mainly preferred in grassland environment. Ideally, it means that once a new cultural tradition has been reconstructed for an important site like Mumba, its cultural entity should be compared to the environment that supported human survival at any particular time. Such comparisons may help enable an assessment of temporal and spatial distribution of change in technology and behaviour, which can then be viewed in the wide context of changing climate to shed light on an interplay between nature and culture.

Results from both TCSA values and use-wear signals have shown good sources for better understanding the technological skills and possible food acquisition strategies employed by MSA foragers at Mumba rock-shelter. It appears that weapons made of stone points were portable and efficient in hunting. They assisted the toolmakers to accumulate reliable food resources, defend themselves, and increase their geographical range inside and outside the African continent during the last phases of the Middle and Late Pleistocene (McBrearty and Brooks, 2000; Mellars, 2006; Willoughby, 2007). Morphologic assessment of points from Mumba MSA assemblages suggests that at least three main design features were in the mind of the toolmakers: a well prepared binding portion that would link the stone tip and organic shaft, elliptical cross-section that make their TCSA values significantly smaller compared to other MSA and Middle Paleolithic points of Eurasia, and a sharp tip that penetrate an animal's body to produce greater wound that could allow bleeding and limit distance in which a prey can travel after it has been shot (Bushozi, 2011, 2013). There seems to be resemblance in TCSA values for points from all three Beds (V, VI-A and VI-B) regardless of variation in time range and environmental dynamics.

Hunting weapons have three basic characteristics: hafting elements, sharp tip, and impact scars (Villa and Lenoir, 2009). Evidence of hafting was found on some Mumba points through micro-wear and butt morphologic appearances (Figure 5). Thinning of the butt by narrowing the original striking platform is generally considered as a basic way to prepare the base of point to meet haft requirements (Villa and Lenoir, 2009; Bushozi, 2011). Thinned butt allows the tie between point as well as shaft and limits protrusion of binding materials from the beam. Use-wear traces indicate that impact fractures resulted from projectile actions predominate, accounting to 26.6%. Although percentage representations are based on a small sample, consistence in projectile elements indicates they were made to meet a specific requirement. Therefore, points could have developed such impact stress signals frequently depending on the rate of launching a projectile to the body of the prey. Alternatively, it may imply that hunting was a primary factor for making stone points. However, to some extent, micro-wear signals indicate that some of the analysed points were not intensively utilized or use signals were removed by depositional or post-depositional processes (Figure 9). The high frequency of points with signals-related impact stress implies that points of Mumba were functionally effective, specifically, for hunting-related activities.

Statistically, TCSA values of stone points from Mumba rock-shelter are significantly smaller than experimentally replicated MSA points that performed better with bow and arrows (Waweru, 2007). Experimental replicates of MSA points with a mean TCSA value of 91.1 mm^2 with SD of 36.7 effectively penetrated a gazelle carcass with bow and arrows (Waweru, 2007). The issue of reduced mean TCSA value for points from Mumba make the issue of weapon systems made and used by MSA forager open for more discussion; but it supports an idea raised by Shea and Sisk in 2010, that projectile weapons have deep antiquity in Sub-Saharan Africa dating as far back as 100 ka. Evidence from this study suggests that sometimes stone points were curated, retooled, and used to perform other activities such as cutting, butchering, and wood and bone processing in their life history; but they made huge contribution to development of projectile technologies. However, we see no reason for preferring one interpretation over the other such that they could have been either depending on the hunting situation, nature of hunted game, or cultural expectations.

In due regard to substantial cultural and technological change over time, the pattern of TCSA values does not appear to have significantly changed. Classifying patterning through microscopic analysis and metric dimensions coupled with graphic representations gives an indication of relative similarities in TCSA values of points from different stratigraphic sequences defined at Mumba regardless of disparity in age settings. The more effective patterns on point production and use show that they were continuously manufactured and utilized in comparable ways. Therefore, it is suggested that MSA foragers responded to cultural and biological dynamics in a closely related manner and in comparable ways through time. Toolmakers were capable of designing different kinds of weapons depending on prey preference and hunting situation. It is possible that the use of spears as well as bow and arrow for hunting practices during the MSA made a huge contribution in human subsistence during the MSA and later cultures. Likewise change of MSA cultural materials at Mumba rock-shelter can be distinguished or confirmed based on archaeological composition, typological classifications, and tool diversity. However, more studies are encouraged before making useful conclusion. It is important that change in sequences at Mumba correlate with other MSA occurrences across the region and quantification of functional variable should also focus on other tools. This could be reached with more emphasis on comparative studies across Africa, in particular, Eastern and Southern Africa where more evidence of human remains, lithic, fauna, and symbolic revealing artefacts have been found in stratified MSA contexts.

10.7 CONCLUSION

The study demonstrates at least four significant changes in depositional environments of Mumba rock-shelter from Beds III to VI-B. They are interpreted as a result of climatic changes from humid conditions during Beds VI-B to V and dry conditions afterwards. Yet, the landscape that surrounds Mumba rock-shelter is characterized by grasslands with few woodlands and shrubs in riverine catchments. Sedimentary deposits of this region are related to volcanism, sedimentological displacements, and tilting or uplifting of the landscape. During this period, especially during Beds VI-B to V about ~130 ka to ~70 ka, anatomically modern humans with MSA technologies lived at Mumba rock shelter. Deep cultural sequence and dense archaeological artefacts suggest that Mumba was used as residential camp and an industrial centre for tool manufacturing and retooling for number of decades. Among material cultures they manufactured were unifacial and bifacial points, but they are represented in low frequency. Preliminary micro-fractures and TSCA value reflect only a minimum number of the lithic artefacts that were used as hunting devices. This is primarily because micro-fractures and use-wear signals on lithic artefacts can often be affected by depositional and post-depositional processes. In addition, hunting devices were potentially shot into game or lost in the field during the hunting practices. The TCSA values, use-wear and micro-fractures show that MSA points were retooled and they performed a wide range of activities such as cutting, butchering as well as wood and bone processing in their life history. Yet, TCSA values and KS test do not show substantial cultural and technological changes over time, suggesting that similar technological traits and hunting skills were inherited and transmitted over generations.

ACKNOWLEDGEMENTS

This research was supported by the Volkswagen Stiftung Foundation Initiative: knowledge for tomorrow cooperative research project in sub-Saharan Africa through a Postdoctoral Fellowship in humanities in Sub-Saharan and northern Africa given to Dr Pastory Bushozi. We sincerely thank the Volkswagen Foundation for funding this study. Thanks are also due to the Antiquities Department, Ministry of Natural Resources and Tourism, Government of Tanzania, and the Tanzania Commission for Science and Technology (COSTECH) for granting research clearance and permits. In the field, we had support of the Village government that we hereby thank. Last but not the least, we thank our graduate students and research assistants.

REFERENCES

Ambrose, S.H., 2001, Middle and Late Age settlement patterns in the Central Rift Valley, Kenya: comparison and contrasts. In *Settlement Dynamics of the Middle Paleolithic and Middle Stone Age*, edited by Conard, N., Tübingen Publication in Prehistory **1**, (Tübingen: Kerns Verlag), pp. 21–43.

Ambrose, S.H., 2002, Small things remembered—Origins and early microlithic tools industries in sub-Saharan Africa. In *Thinking Small: Global perspectives on microlithic toolsization*, edited by Elston, R.G. and Kuhn, S.L., Archaeological Papers of the American Anthropological Association **2**, (Chichester: Wiley), pp. 9–29.

Baker, B.H., Mohr, P.A. and Williams, L.A.J., 1972, Geology of the Eastern Rift System of Africa, *Geological Society of America Special Paper,* **136**, p. 67.

Bretzke, K., Marks, A.E. and Conard, N.J., 2006, Projektiltechnologie und kulturelle Evolution in Ostafrika. *Mitteilungen der Gesellschaft für Urgeschichte*, **15**, pp. 63–81.

Bräuer, G. and Mabulla, A.Z., 1996, A new fossil hominid from Lake Eyasi, Tanzania. *Anthropologie*, **34**, pp. 47–53.

Bräuer, G. and Mehlman, M., 1988, A Hominid Molar from a Middle Stone Age level at Mumba Rock-shelter, Tanzania. *American Journal of Physical Anthropology*, **75**, pp. 69–76.

Brooks, A.S., Hare, P.E., KoKis, J.E., Miller, G.H., Ernst, R.D. and Wendorf, F., 1990, Dating Pleistocene Archeological Sites by Protein Diagenesis in Ostrich Eggshell. *Science*, **248**(4951), pp. 60–64.

Bushozi, P.G.M., 2011, *Lithic technology and hunting behaviour during the MSA in southern and northern Tanzania*. PhD Dissertation, University of Alberta, Edmonton.

Bushozi, P.M., 2013, A functional study of MSA points: evidence from Mumba rock-shelter, Tanzania. *The Journal of African Archaeological Network*, **11**, pp. 25–42.

Cohen, A.S., Stone, J.R., Beuning, K.R.M., Park, L.E., Reinthal, P.N., Christopher, D.D., Scholz, A., Johnson, T.C., King, J.W., Talbot, M.R., Brown, E.T. and Ivory, S.J., 2007, Ecological consequences of early late Pleistocene megadroughts in tropical Africa. *Proceedings of the National Academy of Sciences*, **104**(42), pp. 16422–16427.

Conard, N.J., 2005, An overview of the patterns of behavioral change in Africa and Eurasia during the Middle and Late Pleistocene. In *From Tools to Symbols: From Early Hominids to Modern Humans*, edited by d'Errico, F. and Blackwell, L., (Johannesburg: Witwatersrand University Press), pp. 294–332.

Conard, N.J. and Marks, A., 2006, New research on the MSA of Mumba cave, Tanzania. Paper presented in the 71st Annual Meeting of the Society for American Archaeologists, San Juan.

deMenocal, P.B. and Stringer, C., 2016, *Climate and the peopling of the world*. http://dox.doi.org/10.1038/Nature 19471.

Domínguez-Rodrigo, M., Diez-Martín, F., Mabulla, A., Luque, L., Alcalá, L., Tarriño, A., López-Sanz, J.A., Barba, R. and Bushozi, P., 2007, The archaeology of the Middle Pleistocene deposits of Lake Eyasi (Tanzania). *Journal of African Archaeology*, **5**(1), pp. 47–78.

Domínguez-Rodrigo, M., Mabulla, A., Luque, L., Thomson, K., Rink, J., Diez-Martín, F., Bushozi, P. and Alcalá, L., 2008, A new archaic Homo sapiens fossil from Lake Eyasi, Tanzania. *Journal of Human Evolution*, **54**, pp. 899–903.

Díez-Martín, F., Domínguez-Rodrigo, M., Sánchez, P., Mabulla, A., Tarriño, A., Barba, R., Prendergast, M.E. and Luque, L., 2009, The Middle to Later Stone Age Technological Transition in East Africa. New Data from Mumba Rockshelter Bed V (Tanzania) and their implications for the origin of modern human behavior. *Journal of African Archaeology*, **7**(2), pp. 147–173.

Ebinger, C., Poudjom, Y., Mbede, E.A. and Dawson, J.B., 1997, Rifting Archaean Lithosphere: The Eyasi-Manyara-Natron Rifts. *Journal of the Geological Society*, **154**, pp. 947–960.

Finch, J., Leng, M.J. and Marchant, R., 2009, Late Quaternary vegetation dynamics in a biodiversity hotspot, the Uluguru Mountain of Tanzania. *Quaternary Research*, **72**, pp. 111–122.

Foster, A., Ebinger, C., Mbede, E. and Rex, D., 1997, Tectonic development of the northern Tanzanian sector of the East African Rift System. *Journal of the Geological Society*, **154**(4), pp. 689–100.

Gliganic, L.A., Jacobs, Z., Roberts, R.G., Dominguez-Rodrigo, M., Mabulla, A.Z.P., 2012, *Journal of Human Evolution*, **62**, pp. 533–547.

Grace, R., 1988, *Teach yourself microwear analysis: a guide to interpretation on the function of stone tools*, Arqueohistoria **3**, (Santiago de Compostela: Universidade de Santiago de Compostela).

Grace, R., 1989, *Interpreting the function of stone tools. The quantification and computerization of microwear analysis*, (Oxford: British Archaeological Reports), International Series, **147**.

Hughes, S.S., 1998, Getting to the point: Evolutionary change in prehistoric weaponry. *Journal of Archaeological Method and Theory*, **5**(4), pp. 345–408.

Keeley, L.H., 1982, Hafting and retooling: Effect on the archaeological record. *American Antiquity*, **47**, pp. 98–805.

Kohl-Larsen, L., 1943, *Auf den Spuren des Vormenschen*, (Stuttgart: Strecker und Schröder).

Kohl-Larsen, L. and Kohl-Larsen, M., 1958, *Die Bilderstrasse Ostafrikas*, (Kassel: Erich Roth Verlag).

Lombard, M. and Phillipson, L., 2010, Indications of bow and stone-tipped arrow use 64,000 years ago in KwaZuru-Natal, South Africa. *Antiquity*, **84**, pp. 1–14.

Lombard, M., 2007, Finding resolution for the Howiesons Poort through the microscope: micro-residue analysis of segments from the Sibudu Cave, South Africa. *Journal of Archaeological Science*, **35**(1), pp. 26–41.

Mabulla, A.Z.P., 1996, *Middle and Later Stone Age land use and lithic technology in the Eyasi Basin, Tanzania*. PhD Dissertation, University of Florida.

Mabulla A.Z.P., 2007, Hunting and foraging in the Eyasi Basin, northern Tanzania: past, present and future prospectus. *African Archaeological Review*, **24**, pp. 15–33.

Manega, P.C., 1993, *Geochronology, Geochemistry and Isotopic Study of the Plio-Pleistocene Hominid Sites and the Ngorongoro Volcanic Highlands in Northern Tanzania*. PhD Dissertation, University of Colorado, Boulder.

McBrearty, S. and Brooks, A.S., 2000, The revolution that wasn't: a new interpretation of origin of modern behavior. *Journal of Human Evolution*, **30**, pp. 452–565.

Mehlman, M., 1989, *Later Quaternary Archaeological sequences in Northern Tanzania*. PhD Dissertation, University of Illinois, Urbana Champaign.

Mellars, P., 2006, Why did modern human populations disperse from Africa 60,000 years ago? A new model. *Proceedings of the National Academy of Sciences*, **103**(25), pp. 9381–9386.

Merrick, V.H. and Brown, F.H., 1984, Obsidian sources and patterns of resource utilization in Kenya and northern Tanzania: Some initial findings. *African Archaeological Review*, **2**, pp. 129–152.

Mohapi, M., 2007, Rose Cottage Cave MSA lithic points: Does technological change imply change in hunting techniques? *South African Archaeological Bulletin*, **62**(185), pp. 9–18.

Pickering, R., 1961, The geology of the country around Endulen. *Records of the Geological Survey of Tanganyika*, **XI**, pp. 1–9.

Prendergast, M.E., Luque, L., Dominguez-Rodrigo, M., Martin, F.D., Mabulla, A.Z., Barba, R., 2007, New excavation at Mumbarock-shelter, Tanzania. *Journal of African Archaeology*, **5**(2), pp. 217–243.

Protsch, R., 1981, The Kohl-Larsen Eyasi and Garusi hominid finds in Tanzania and their relation to Homo erectus, In *Homo erectus: papers in honor of Davidson Black*, edited by Sigmon, B.A. and Cybulski, J.S., (Toronto: Toronto University Press), pp. 217–226.

Rafalski, S., Schröter, P. and Wagner, E., 1978, Die Funde am Eyasi-Nordostufer. In *Die archäologischen und anthropologischen Ergebnisse der Expeditionen in Nord-Tanzania, 1933–1939*, edited by Müller-Beck, H., Tübinger Monographien zur Urgeschichte, **4**(2).

Reimer, P.J., Baillie, M.G.L., Bard, E., Bayliss, A., Beck, J.W., Bertrand, C.J.H., Backwell, P.G., Buck, C.F., Burr, G.S., Cutler, K.B., Damon, P.E., Edwards, R.L., Fairbanks, R.G., Friedrich, M., Guilderson, T.P., Hogg, A.G., Hughen, K.A., Kromer, B., McCormac, F.G., Manning, S.W., Ramsey, C.B., Reimer, R.W., Remmele, S., Southon, J.R., Stuiver, M., Talamo, S., Taylor, F.W., Van der Plicht, J. and Weyhenmeyer, C.E., 2004, Terrestrial radiocarbon age calibration, 26–0 ka BP. *Radiocarbon*, **46**, pp. 1029–1058.

Sano, K., 2009, Hunting evidence from stone artefacts from the Magdalenian cave site Bois Laiterie, Belgium: a fracture analysis. *Quartär*, **56**, pp. 67–86.

Scholz, C.A., Johnson, T.C., Cohen, C.S., King, J.W., Peck, J.A., Overpeck, J.T., Talbot, M.R., Brown, E.T., Kalindekafe, L., Amoako, P.Y.O., Lyons, R.P., Shanahan, T.M., Castaneda, I.S., Heil, C.W., Forman, S.L., McHargue, L.R., Beuning, K.R., Gomez, J., and Pierson, J., 2007, East African megadroughts between 135 and 75 thousand years ago and bearing on early modern origins. *Proceedings of the National Academy of Sciences*, **104**(42), pp. 16416–16421.

Shea, J. and Sisk, M.L., 2010, Complex projectile technology and Homo sapiens dispersal into Western Eurasia. *Paleoanthropology Society*, **10**(36), pp. 100–122.

Shea, J., 2006, The origins of lithic projectile point technology: evidence from Africa, the Levant and Europe. *Journal of Archaeological Science*, **20**(3005), pp. 1–4.

Shea, J., 2009, The impact of projectile weaponry on the Late Pleistocene hominid evolution, In *The evolution of human diets: Intergraded approaches to the study of Palaeolithic subsistence*, edited by Hublin, J.J. and Richard, M.P., (New York: Springer), pp. 189–199.

Sisk, M. and Shea, J., 2011, The African origin of complex technology: an analysis using Tip Cross-Section Area and Perimeter. *International Journal of Evolutionary Biology*, **1**, pp. 1–8.

Skinner, A.R., Hay, R.L., Masao, F. and Blackwell, B.A., 2003, Dating the Naisiusiu Beds, Olduvai Gorge by electron spin resonance. *Quaternary Geochronology*, **22**, pp. 1361–1366.

Tierney, J.E., Russell, J.M., Damsté, J.S.S., Huang, Y. and Verschuren, D., 2011, Late Quaternary behaviour of the East African monsoon and the importance of the Congo Air Boundary. *Quaternary Science*, **30**, pp. 798–807.

Thomas, D.H., 1978, Arrowheads and Darts—How the stones got to the shaft. *American Antiquity*, **43**(3), pp. 461–472.

Villa, P. and Lenoir, M., 2009, Hunting and hunting weapons of Lower and Middle Paleolithic of Europe. In *The evolution of human diets: Intergraded approaches to the study of Palaeolithic subsistence*, edited by Hublin, J.J. and Richard, M.P., (New York: Springer), pp. 189–199.

Waweru, V.J., 2007, *Middle Stone Age technology at Cartwright's site, Kenya*. PhD thesis. University of Connecticut. Connecticut.

Wilkins et al., 2012, Evidence for early hunting technology. *Science Magazine*, **338**(6109), pp. 942–946.

Willoughby, P.R., 2007, *The evolution of modern human in Africa: A comprehensive guide*, (Lanham: Altamira Press).

Willoughby, P.R., 2012, The Middle and Later Stone Age in the Iringa Region of southern Tanzania. *Quaternary International*, **270**, pp. 103–118.

Wurz, S. and Lombard, M., 2007, 70,000 years old geometric backed tools from the Howiesons Poort at Klasies River, South Africa—Were they used for hunting? *South African Humanities*, **19**, pp. 1–16.

CHAPTER 11

Spatial distribution and impacts of mining Development Minerals in Greater Accra Metropolitan Area, Ghana

Rosemary Okla

Ghana Geological Survey Authority, Accra, Ghana

ABSTRACT: Natural resources development is the biggest asset for rapid and sustainable development of a country. However, information on natural resources available in Africa is much less than in other continents in the world. This and the lack of sufficient funds for geological mapping are affecting Africa's ability to maximize its potential of resources use. Therefore, in most African countries, there is the need to provide transparent public geological information. In recent times, such information is acquired by exploration of precious minerals such as gold, diamond and silver. However, such explorations are not carried out at all or not to an equal degree for Development Minerals or industrial minerals like construction material, dimension stones, and semi-precious stones. Development Minerals, especially those extensively used as bulk construction materials e.g. sand and quarry stones, stimulate the rapid transformation of cities and of many metropolitan and urban areas in developing countries. The phenomenal growth recently in infrastructure such as housing, office complexes and road construction have all benefited from extensive exploitation of such materials, which are transported and used in the construction and allied industries from mainly peri-urban and rural communities. Exploitation of these materials provides employment, livelihoods, and income to many low-income households and women, in particular. In spite of their increasingly socio-economic importance, however, not much research has been done to systematically map and document the spatial distribution of such materials to know their likely effect over time on the physical landscape or on the health and safety of people directly engaged in mining and on communities that are close to these mining sites.

This project seeks to identify Development Minerals that are extracted in some parts of Greater Accra Metropolitan Area, by localizing the mining sites of these minerals and representing them in a map using a Geographic Information System (GIS), and observing and identifying the impact of the extraction on the affected people. This approach provides relevant data for policy planning and implementation, such as maximizing the benefits derived from the exploitation and use of Development Minerals in the metropolitan areas in Ghana as well as assessing the possible negative impacts of such exploitations and updating Ghana's geoscientific data.

11.1 INTRODUCTION

Natural resources have a direct impact on a country's economy. The fastest-growing countries in the world are resource-rich countries like China, India and Brazil, making mineral resources indispensable to economic growth in this century (McMahon and Moreira, 2014). Since mineral resources are a vital commodity for rapid development of a nation, most countries research on their availability. However, in Africa there is still a significant lack of such research carried out by the public sector (Souza and Mogessie, 2016), and therefore, there is a need to provide transparent public geological information on mineral resources.

According to Kesse (1985) over 2000 minerals are known in Ghana but only about 100 of these minerals are common. This proves that more research should be done to identify the various mineral resources. In recent times, much geological information has been available on precious minerals such as gold, diamond, and silver, but little or no information on Development Minerals or industrial minerals like construction materials, dimension stones, and semi-precious stones is available due to the less fiscal returns.

Development Minerals are useful for economic, social, and human development. They are of low value to the international commodity market but rather provide essential domestic commodity for construction, infrastructure manufacturing, etc. According to Franks (2015) Development Minerals are "the hidden bedrock of our society" helping nations to attain sustainable development and creating employment opportunities to boost the livelihoods of its citizens.

11.1.1 Economic contribution of precious and Development Minerals

Comparing the economic contribution of precious minerals to Development Minerals in the area of foreign direct investment (FDI), it is high in precious minerals and low in Development Minerals. Governments generate 3–10% national income from precious minerals and less from Development Minerals. However, Development Minerals contribute largely to domestic value addition by feeding local industries. Regarding employment, only a small percentage of about 1–2% of the total employment in a country can be accredited to Precious Minerals mining sector; whereas in Development Minerals sector, larger number of unskilled labour are employed (ICMM, 2012).

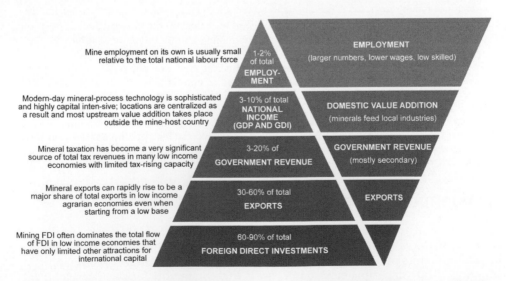

Figure 1. Diagram showing the comparison between precious minerals (left) and Development Minerals (right) (modified after ICMM, 2012).

In the case of Ghana, precious minerals contribute to 50% of foreign direct investment and 5% of the gross domestic product into the economy, with only 0.7% employed in the precious minerals mining sector (Ghana Statistical Services, 2008; Baah, 2005; UNCTAD, 2005).

Precious and Development Minerals contribute to a nation's economy in diverse ways (Figure 1). Therefore, if governments pay more attention to Development Minerals, resource-rich countries will not depend only on precious minerals for revenues only, but also on Development Minerals to diversify their economies.

This paper aims to map the spatial distribution of Development Minerals and the conceptual impacts of Development Minerals at some locations of Greater Accra Metropolitan Area (GAMA), Ghana using Geographic Information Systems (GIS). Maps are a useful medium to communicate such information, especially to policy makers. Access to the spatial distribution of Development Minerals in GAMA enhances easy communication among stakeholders and better trade among investors.

11.1.2 Development Minerals in Greater Accra Metropolitan Area and its uses

The Development Minerals found in the GAMA are construction materials like sand and gravel as well as industrial minerals such as salt, gypsum, and dimension stones like boulders of slate (Kesse, 1985). Construction materials are aggregates produced from bedrocks used in their natural form like boulders, pebbles, gravels, sands, and clays (Kesse, 1985). Particularly, 97% of sand and gravel is consumed in the construction industry (Goldman, 1994) as concrete for bridges, apartment structures, roads and others such infrastructure.

Industrial minerals are "any rock, mineral or other naturally occurring substance of economic value, exclusive of metal ores, mineral fuels, and gemstones" (Bates, 1975). Salt and gypsum are examples of industrial minerals where salt has more than 14000 uses (Kesse, 1985), which include, among others, use of salt as an additive in chemical feedstock, to soften and condition water, in fertilizers and insecticides, in the manufacture of chlorine and caustic soda, etc. Gypsum is mainly used in the building industry as binding material (Kesse, 1985).

Dimension stone is a natural rock material quarried for the purpose of obtaining blocks or slabs that meet specifications as to size and shape (Barton, 1968). It is used as building material.

Mapping Development Minerals helps improve their contribution to the Gross National Product (GNP). Maps offer a quick way to interpret information on mineral resources with less assistance and also to comprehend spatial relationships between the occurrence of Development Minerals and the underlying geology. An efficient way to produce communicative maps is the use of Geographic Information Systems (GIS). GIS help visualize, analyse, and interpret data; and to understand the spatial relationships between the data (Clarke, 1986). GIS was used to visualize the spatial distribution of the mapped Development Minerals in GAMA. The outcome of this project has helped update Ghana's geoscientific data, which has been brought to the attention of the public through presentations to stakeholders.

11.2 STUDY AREA

The Greater Accra Metropolitan Area (GAMA) was chosen as study area because it is characterized by emerging infrastructure, construction sites, and industries, which have high demands for Development Minerals.

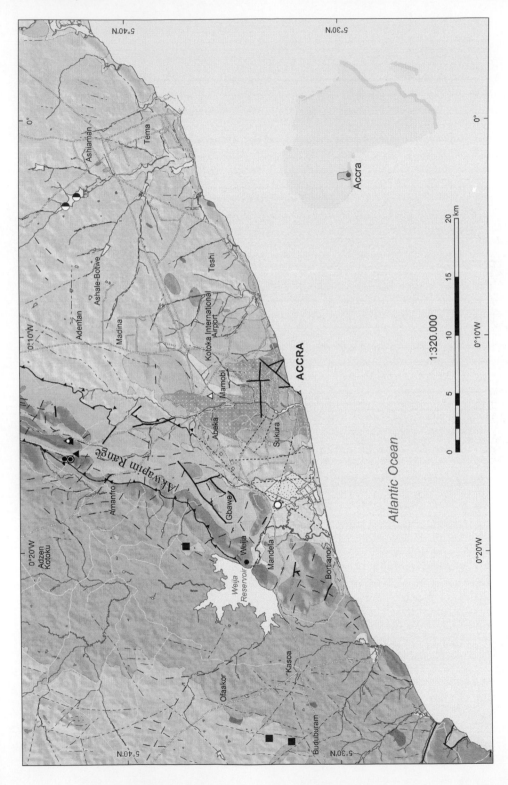

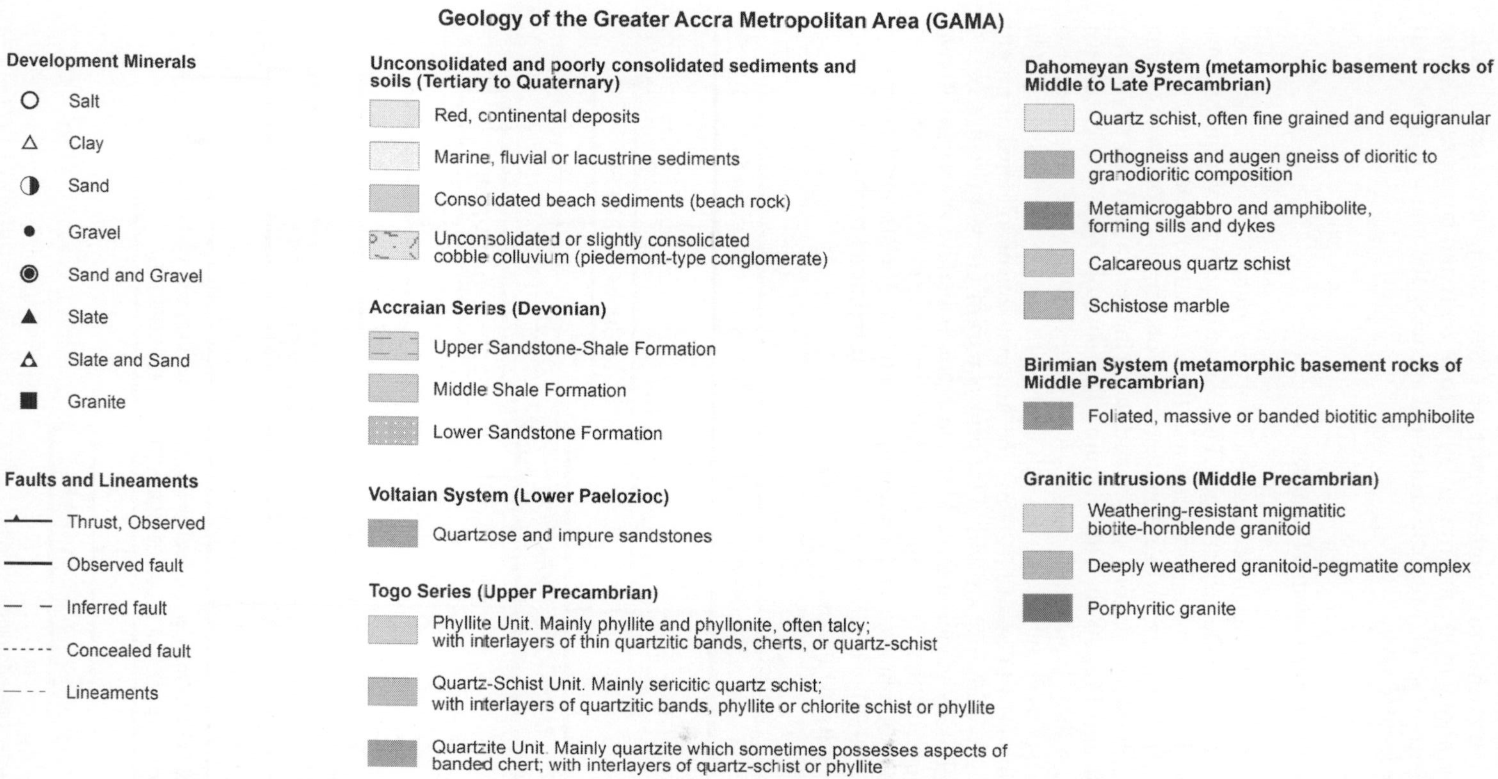

Geology of the Greater Accra Metropolitan Area (GAMA)

Development Minerals

○ Salt

△ Clay

◐ Sand

● Gravel

◉ Sand and Gravel

▲ Slate

△ Slate and Sand

■ Granite

Faults and Lineaments

Thrust, Observed

Observed fault

Inferred fault

Concealed fault

Lineaments

Unconsolidated and poorly consolidated sediments and soils (Tertiary to Quaternary)

Red, continental deposits

Marine, fluvial or lacustrine sediments

Consolidated beach sediments (beach rock)

Unconsolidated or slightly consolidated cobble colluvium (piedemont-type conglomerate)

Accraian Series (Devonian)

Upper Sandstone-Shale Formation

Middle Shale Formation

Lower Sandstone Formation

Voltaian System (Lower Paeleozioc)

Quartzose and impure sandstones

Togo Series (Upper Precambrian)

Phyllite Unit. Mainly phyllite and phyllonite, often talcy; with interlayers of thin quartzitic bands, cherts, or quartz-schist

Quartz-Schist Unit. Mainly sericitic quartz schist; with interlayers of quartzitic bands, phyllite or chlorite schist or phyllite

Quartzite Unit. Mainly quartzite which sometimes possesses aspects of banded chert; with interlayers of quartz-schist or phyllite

Dahomeyan System (metamorphic basement rocks of Middle to Late Precambrian)

Quartz schist, often fine grained and equigranular

Orthogneiss and augen gneiss of dioritic to granodioritic composition

Metamicrogabbro and amphibolite, forming sills and dykes

Calcareous quartz schist

Schistose marble

Birimian System (metamorphic basement rocks of Middle Precambrian)

Foliated, massive or banded biotitic amphibolite

Granitic intrusions (Middle Precambrian)

Weathering-resistant migmatitic biotite-hornblende granitoid

Deeply weathered granitoid-pegmatite complex

Porphyritic granite

Figure 2. Geological map of the GAMA study area, modified from Geological Survey Department (2006); Draft: R. Okla, Cartography: J. Eisenberg. Disclaimer: Every effort has been made to accurately collect, create and represent the information and data shown on this map. The responsibles and contributors have made an extensive effort to assess the quality of the data entered for consistency and accuracy. The use of the information is solely at the risk and option of the user.

The GAMA lies within the southern part of Ghana along the coastal plain. It covers an area of about 1400 km² from 0° 30′W to 0° 05′E and stretches from the coastline at 5° 30′N to 5° 45′N. Most governmental agencies, the major seaport, and industrial areas are located here.

11.2.1 Geological and geomorphological features

The study area is underlain by heterogeneous rocks of Precambrian, Lower Paleozoic, Devonian, and Tertiary up to Quaternary ages. The Precambrian rocks within this area are the Togo Series of Upper Precambrian age and the Birimian and Dahomeyan systems of Middle to Late Precambrian age. The Birimian, Voltaian, and Togo rocks occur mainly in the North Western part of GAMA and the Dahomeyan rocks in the North Eastern part (Muff and Efa, 2006). The metamorphic rocks of the Togo Series have been subdivided into three main units, which are phyllite or phyllonite, quartz-schist, and quartzite, which consists of siliciclastic sediments with low- to intermediate grade type of metamorphism (Muff and Efa, 2006). The Accraian, unconsolidated or poorly consolidated sediments and soils, are within the central part of GAMA (Figure 2).

The geomorphology is characterised by the NNE-extending Akwapim Range and the Weija Mountains that fall within the Togo Series (Figure 2). The escarpment has a height of up to 365 m a.s.l. and a width of 3 to 6 km. The structure is striking from NE to SW and dipping to SE and ESE directions. It is limited by the western and eastern boundary faults (Muff and Efa, 2006).

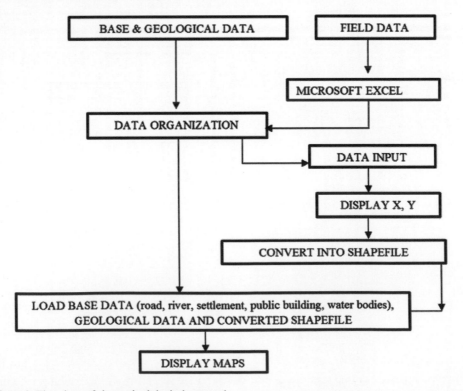

Figure 3. Flowchart of the methodological approach.

11.3 METHODOLOGY

The base and geological data used was acquired from the Geological Survey Department in the Ghana-Germany Environmental and Engineering Geology Project at a scale of 1:10,000 and 1:100,000 (Muff and Efa, 2006) in Ghana National Grid projection. Information on Development Minerals and its uses were collected from reports from Geological Survey Department of Ghana (2006), Mineral Commission and from Barton (1968), Bates (1975), and Goldman (1994). The locations of Development Minerals that were identified from the reports were collected in the field with a handheld GPS. Also, the environmental and social impact of mining by Development Minerals was observed.

The field data was organized in Microsoft Excel. The excel file was imported in ArcMap 10.1 (ESRI) according to the Latitude/Longitude specifications of the handheld device and transformed to a shapefile afterwards. The shapefile was projected to Ghana National Grid. Additionally, the base data and the geological data were added to the GIS project (Figure 3).

11.4 RESULTS

The overview map (Figure 2) shows the collected locations of Development Minerals in the study area. Most of the minerals fall within the hilly area of the Akwapim Range and the Weija Mountains belonging to the stratigraphic unit of the Togo Series (Figure 2). It was also observed that some of the construction materials and dimension stones are granitoidic intrusions of Middle Precambrian age. The steep dipping rocks are foliated striking from N to NNE direction (Muff and Efa, 2006). According to Muff and Efa (2006) "the granitoids are very susceptible to mechanical weathering, which causes the disintegration of the bedrock and the formation of a several meter thick surface layer of disintegrated quartz and feldspar grains. They are an important source for construction sand."

Despite the fact that Development Minerals contribute to the secondary revenue of a country, mining these minerals often leave a negative impact to the affected communities. According to the World Bank (2002) the mining industry is a "footprint industry," which means it leaves an environmental, social, and economic impact wherever it is active. The environmental impacts of these mining activities have diverse and devastating effect on the environment as well as the people who live in the mined communities because basic elements such as land, water, and air are affected by these activities (Amponsah and Dartey, 2011).

During field work environmental impacts of mining activities were observed at the quarry sites. At the mining sites, activities such as drilling, blasting, excavating, crushing operation, and pile storage of sand and gravel generate noise and dust particles that are harmful to both workers and surrounding communities, which can cause health hazards and respiratory diseases such as asthma. At most sites, workers do not wear personal protective equipment (Figure 4). As per the Inspectorate Division of the Minerals Commission, the common health problems by mining activities from 2000–2004 include malaria and upper respiratory tract infection (Amponsah and Dartey, 2011; Ghana Health Service, 2007).

The dust pollutes the surface water which compromises the water quality, thus affecting its aquatic life, wildlife and other uses of the surface and groundwater (Figure 5). Removal of top soil and its overburdening has caused damage to the landscape. Also, the excavated sites are left open and filled with rain water, which serves as breeding ground for mosquitoes (Figure 5).

Figure 4. Stakeholder participation: Interviewing a woman at a quarry site near Wejia crushing stones. The woman is not wearing a protective gadget.

Figure 5. A former quarry site with modified landscape re-filled partially by water which serves as a breeding ground for mosquitoes.

11.5 CONCLUSION

Mapping the spatial distribution of Development Minerals will help developers in easily locating and accessing construction materials in the GAMA besides helping update Ghana's geoscientific data. Six Development Minerals were identified, which are clay, granite, salt, slate, sand, and gravel. Even though gypsum occurs in the area, it has no economic occurrences and was therefore not considered.

The increasing need for materials for the construction industry has given rise to high demand of these materials (Jafaru, 2013). There is also a high demand of salt in the West African sub-region (Affam and Asamoah, 2011). These attest the fact that the policy makers should place more emphasis on Development Minerals and promote this sector.

All the quarry sites fall within rocks of Middle Precambrian age, specifically the Togo series. It was shown that the weathered granitoids are an important source of construction sand (Muff and Efa, 2006).

The mining policy of Ghana sees to it that all mining activities, while at the same time, the negative impacts on the environment, the workers and the general public are minimized. The only way to achieve this is through compliance (Ministry of Lands and Natural Resources, 2016). As gathered from field observations, some quarry sites adhere to the Environmental Protection Agency's policies; whereas a few affect the environment, the workers and the general public in the surroundings of the mining site negatively. It was observed that some quarry sites used the manual operation, which impedes productivity and does not meet the high demand of materials.

REFERENCES

Affam, M. and Asamoah, D., 2011, Economic Potential of Salt Mining in Ghana Towards the Oil Find. *Research Journal of Environmental and Earth Sciences,* **3**(5), pp. 448–456.

Amponsah, T. and Dartey, B., 2011, The Mining Industry in Ghana: A Blessing or a Curse. *International Journal of Business and Social Science*, **2**(12), pp. 62–69.

Baah, A., 2005, Assessing labour and environmental standards in South African multinational companies in the mining industry in Africa: The case of Goldfields South Africa. In *Mining Africa, comprehensive report of South African MNCs labour and social performance,* edited by Pillay, D., (South Africa: National Labour and Economic Development Institute).

Barton, W., 1968, *Dimension stone.* U.S. Bureau of Mines Information Circular, **8391**, pp. 147.

Bates, R., 1975, *Industrial Minerals and Rocks* (4th ed.), edited by Lefond S.J., (New York: American Institute of Mining, Metallurgical, and Petroleum Engineers).

Clarke, K., 1986, Advances in geographic information systems, computer, environment and urban systems, *Journal of Computer, Environment and Urban Systems*, **10**, pp. 175–184.

Franks, D., 2015, What has salt got to do with development? *Our Perspectives.* Blog of the United Nations Development Programme. – 22.09.2016: http://www.undp.org/content/undp/en/home/blog/2015/11/23/What-has-salt-got-to-do-with-development-.html.

Geological Survey Department, 2006, Engineering Geology Map of Greater Accra Metropolitan Area, Geological Map for Urban Planning 1:100,000; Accra.

Ghana Health Service, 2007, *Ghana Health Service Annual Report*, (Accra: Government of Ghana).

Ghana Statistical Services, 2008, *Ghana Living Standards Survey Report of the fifth round* (GLSS **5**). – 10.05.2017: http://www.statsghana.gov.gh/docfiles/ glss5_report.pdf.

Goldman, H.B., 1994, Sand and Gravel. In *Industrial Minerals and rocks*, edited by Carr, D.D., 6th Edition, (Littleton: Society for mining, metallurgy and exploration, Inc.), pp. 869–877.

ICMM–International Council on Mining and Metals, 2012, *The role of mining in national economies.* – 22.09.2016: http://www.icmm.com/publications/pdfs/4440.pdf.

Jafaru, M., 2013, *Assessing the socio-economic and ecological impacts of gravel mining in the Savelugu-Nanton District of the Northern Region of Ghana*, (Kumasi: Kwame Nkrumah University of Science and Technology).

Kesse, G.O., 1985, *The Mineral and Rock Resources of Ghana*, (Rotterdam: A.A. Balkema).

McMahon, G.J.R. and Moreira, S., 2014, *The Contribution of the Mining Sector to Socioeconomic and Human Development.* Extractive Industries for Development Series **30** (World Bank, Oil, Gas, and Mining Unit Working Paper).

Ministry of Lands and Natural Resources, 2016, *Minerals and Mining Policy of Ghana.* (Accra: Government of Ghana).

Muff, R. and Efa, E., 2006, Ghana–Germany Technical Cooperation Project: Environmental and Engineering Geology for Urban Planning in the Accra-Tema Area. Accra.

Souza, K. and Mogessie, A., 2016, *African Geological and Mineral Information System (GMIS) Strategy—Promoting geological knowledge as a tool for governance.* (Addis Ababa: United Nations Economical Commission for Africa).

UNCTAD, 2005, *World Investment Report: Transitional Cooperation, Extractive Industries and Development.* (New York & Geneva: United Nations).

World Bank–Group Mining Department, 2002, *Global Mining. Treasure or Trouble? Mining in Developing Countries,* Mining and Development, (Washington: International Finance Cooperation).

Regional Index

Subject Index

Palaeoecology of Africa

International Yearbook of Landscape Evolution and Palaeoenvironments

ISSN: 2372-5907

Volumes 1-12 *Out of Print*

13. Palaeoecology of Africa and the Surrounding Islands
    Editors: J.A. Coetzee & E.M. van Zinderen Bakker
    1981, ISBN: 978-90-6191-203-3

14. Palaeoecology of Africa and the Surrounding Islands
    Editors: J.A. Coetzee & E.M. van Zinderen Bakker
    1982, ISBN: 978-90-6191-204-0

15. Palaeoecology of Africa and the Surrounding Islands
    Editors: J.A. Coetzee, E.M. van Zinderen Bakker, J.C. Vogel,
    E.A. Voigt & T.C. Partridge
    1982, ISBN: 978-90-6191-257-6

16. Palaeoecology of Africa and the Surrounding Islands
    Editors: J.A. Coetzee & E.M. van Zinderen Bakker
    1984, ISBN: 978-90-6191-510-2

17. Palaeoecology of Africa and the Surrounding Islands
    Editors: J.A. Coetzee & E.M. van Zinderen Bakker
    1986, ISBN: 978-90-6191-625-3

18. Palaeoecology of Africa and the Surrounding Islands
    Editor: K. Heine
    1987, ISBN: 978-90-6191-689-5

19. Palaeoecology of Africa and the Surrounding Islands – *Out of Print*
    Editors: K. Heine & J.A. Coetzee
    1988, ISBN: 978-90-6191-834-9

20. Palaeoecology of Africa and the Surrounding Islands
    Editor: K. Heine
    1989, ISBN: 978-90-6191-880-6

21. Palaeoecology of Africa and the Surrounding Islands – *Out of Print*
    Editors: K. Heine & R.R. Maud
    1990, ISBN: 978-90-6191-997-1

22. Palaeoecology of Africa and the Surrounding Islands
    Editors: K. Heine, A. Ballouche & J. Maley
    1991, ISBN: 978-90-5410-110-9

23.   Palaeoecology of Africa and the Surrounding Islands
      Editor: K. Heine
      1993, ISBN: 978-90-5410-154-3

24.   Palaeoecology of Africa and the Surrounding Islands
      Editor: K. Heine
      1996, ISBN: 978-90-5410-662-3

25.   Palaeoecology of Africa and the Surrounding Islands – *Out of Print*
      Editors: K. Heine, H. Faure & A. Singhvi
      1999, ISBN: 978-90-5410-451-3

26.   Palaeoecology of Africa and the Surrounding Islands
      Editors: K. Heine, L. Scott, A. Cadman & R. Verhoeven
      1999, ISBN: 978-90-5410-476-6

27.   Palaeoecology of Africa and the Surrounding Islands: Proceedings
      of the 25th INQUA Conference, Durban, South Africa, 3–11 August 1999
      Editors: K. Heine & J. Runge
      2001, ISBN: 978-90-5809-350-9

28.   Dynamics of Forest Ecosystems in Central Africa during the Holocene:
      Past – Present – Future
      Editor: J. Runge
      2007, ISBN: 978-0-415-42617-6

29.   Holocene Palaeoenvironmental History of the Central Sahara
      Editors: R. Baumhauer & J. Runge
      2009, ISBN: 978-0-415-48256-1

30.   African Palaeoenvironments and Geomorphic Landscape Evolution
      Editor: J. Runge
      2010, ISBN: 978-0-415-58789-1

31.   Landscape Evolution, Neotectonics and Quaternary Environmental Change in
      Southern Cameroon
      Editor: J. Runge
      2012, ISBN: 978-0-415-67735-6

32.   New Studies on Former and Recent Landscape Changes in Africa
      Editor: J. Runge
      2014, ISBN: 978-1-138-00116-9

33.   Changing Climates, Ecosystems and Environments within Arid
      Southern Africa and Adjoining Regions
      Editor: J. Runge
      2016, ISBN: 978-1-138-02704-6

Printed and bound by CPI Group (UK) Ltd, Croydon, CR0 4YY

25/10/2024

01779130-0001